SPORTS AND CITY MARKETING IN EUROPEAN CITIES

T0251903

The European Institute for Comparative Urban Research, EURICUR, was founded in 1988 and has its seat with Erasmus University Rotterdam. EURICUR is the heart and pulse of an extensive network of European cities and universities. EURICUR's principal objective is to stimulate fundamental international comparative research into matters that are of interest to cities. To that end, EURICUR co-ordinates, initiates and carries out studies of subjects of strategic value for urban management today and in the future. Through its network EURICUR has privileged access to crucial information regarding urban development in Europe and North America and to key persons at all levels, working in different public and private organisations active in metropolitan areas. EURICUR closely cooperates with the Eurocities Association, representing more than 100 large European cities.

As a scientific institution, one of EURICUR's core activities is to respond to the increasing need for information that broadens and deepens the insight into the complex process of urban development, among others by disseminating the results of its investigations by international book publications. These publications are especially valuable for city governments, supra-national, national and regional authorities, chambers of commerce, real estate developers and investors, academics and students, and others with an interest in urban affairs.

Euricur website: http://www.euricur.nl

This book is one of a series to be published by Ashgate under the auspices of EURICUR, European Institute for Comparative Urban Research, Erasmus University, Rotterdam. Titles in the series are:

Urban Tourism
Leo van den Berg, Jan van der Borg and Jan van der Meer

Metropolitan Organising Capacity
Leo van den Berg, Erik Braun and Jan van der Meer

National Urban Policies in the European Union
Leo van den Berg, Erik Braun and Jan van der Meer

The European High-Speed Train and Urban Development
Leo van den Berg and Peter Pol

Growth Clusters in European Metropolitan Cities
Leo van den Berg, Erik Braun and Willem van Winden

Information and Communications Technology as Potential Catalyst for Sustainable Urban Development
Leo van den Berg and Willem van Winden

Sports and City Marketing in European Cities

LEO VAN DEN BERG
ERIK BRAUN
ALEXANDER H.J. OTGAAR

Euricur

EUROPEAN INSTITUTE FOR COMPARATIVE URBAN RESEARCH

European Institute for Comparative Urban Research
Erasmus University Rotterdam
The Netherlands
www.euricur.nl

Routledge
Taylor & Francis Group

LONDON AND NEW YORK

First published 2002 by Ashgate Publishing

Published 2016 by Routledge
2 Park Square, Milton Park, Abingdon, Oxfordshire OX14 4RN
711 Third Avenue, New York, NY 10017, USA

First issued in paperback 2016

Routledge is an imprint of the Taylor & Francis Group, an informa business

Copyright © Leo van den Berg, Erik Braun and Alexander H.J. Otgaar 2002

All rights reserved. No part of this book may be reprinted or reproduced or utilised in any form or by any electronic, mechanical, or other means, now known or hereafter invented, including photocopying and recording, or in any information storage or retrieval system, without permission in writing from the publishers.

Notice:
Product or corporate names may be trademarks or registered trademarks, and are used only for identification and explanation without intent to infringe.

British Library Cataloguing in Publication Data
Berg, Leo van den, 1948-
    Sports and city marketing in European cities. - (EURICUR series)
    1. City promotion - Europe 2. Sports - Economic aspects - Europe 3. City planning - Europe
    I. Title II. Braun, Erik III. Otgaar, Alexander H.J.
    IV. EURICUR
    307.1' 4' 164'094

Library of Congress Control Number: 2002101561

ISBN: 978-1-138-25482-4 (pbk)
ISBN: 978-0-7546-1931-4 (hbk)

Contents

List of Figures

List of Tables

Preface

In a globalising and commercialising world, the impact of sports events on our society is increasing day by day. As a consequence, cities more and more become aware of the potential benefits they may derive from acquiring and organising such events. Indeed, sports (clubs, venues and events) can be an instrument to reach objectives of urban management.

This publication aims to analyse the conditions to realise synergies between sports and city marketing, by investigating the experiences of Barcelona, Helsinki, Manchester, Rotterdam and Turin. The analysis is the result of a research project carried out by the European Institute for Comparative Urban Research (EURICUR) from the Erasmus University of Rotterdam, on the invitation of the city of Rotterdam.

For the Rotterdam case study we would like to thank Mr Hans Zoethoutmaar and Ms Karin Luyendijk (Rotterdam Topsport), Mr Marco Roomer (Rotterdam Leisure Department), Mr Karel Mulder (Bestuursdienst; City of Rotterdam) and Ms Marjoleine van Doorn-Claassen (Rotterdam City Development Corporation) for their help. With regard to the four other case studies, we would like to thank Mr Oriol Nel.lo and Mr Louis van Gaal (Barcelona), Mr Eero Holstilla and Mr Timo Cantell (Helsinki), Mr Dave Carter (Manchester) and Ms Ilda Curti and Mr Paolo Bellino (Turin) for organising our visits and establishing contacts with interesting discussion partners. Furthermore, we render thanks to the discussion partners (see appendix) who were very co-operative in sharing their knowledge with us.

Finally, we would like to thank Ms Ankimon Vernède for her unconditional support from the EURICUR secretariat, and Ms Attie Elderson-De Boer for checking our use of the English language.

Leo van den Berg
Erik Braun
Alexander H.J. Otgaar

Chapter One

Sports and City Marketing: An Introduction

Introduction

From 10 June to 2 July, the EURO 2000 football championships took place in Belgium and the Netherlands. The city of Rotterdam had the pleasure of hosting five matches, including the finals. Football supporters from all over Europe visited the city and television viewers all over the world watched these games. The event thus helped to promote the city as a tourist destination and business location. In other words, it was an instrument of city marketing.

This book takes a closer look at the 'state of the art' or sports and city marketing in five European cities, including Rotterdam. The research is meant to increase insight into the role of sports in city marketing and to provide guidelines for an integrated sports and city marketing policy. In other words, to find out how sports (accommodations, clubs and events) can be made an instrument of city marketing and how cities can maximise their potential through a sports and city marketing policy.

The first chapter presents the research framework. In the next section, we will discuss some of the major trends in urban Europe that are inducing European cities to be more competitive and entrepreneurial. Thereafter, we focus on the principles of city marketing as an instrument of urban management, followed by a discussion of the social, economic and cultural values of sports. The next step is to bring sports and city marketing together and assess the potential role of sports in city marketing. The confrontation between sports and city marketing results in a schematic overview of the research framework, which has functioned as a guideline for the case studies. The chapter ends with a short summary of the research approach.

The five subsequent chapters correspond to the five case studies: Barcelona, Helsinki, Manchester, Rotterdam and Turin. These cities can be characterised as sport-minded. Barcelona used the 1992 Olympic Games as a catalyst for the regeneration of its city centre; FC Barcelona is one of the world's most famous football clubs. Helsinki organised the Summer Olympics in 1952 and has run for the Winter Olympics of 2006; the city has a good reputation for organising sports events. Manchester will host the Commonwealth Games in 2002; the city is, of course, known all over the world as the home of Manchester United (although the stadium is in Trafford). Rotterdam was the host for the finals of EURO 2000. Finally, Turin will be the host city of the Winter Olympics in 2006 (the city defeated Helsinki, among others) and the home city of Juventus.

The last chapter presents the most important conclusions related to the research framework to be introduced in the following sections.

Urban Dynamics in Europe

Before we discuss sports and city marketing, it is necessary to put it in the wider perspective of urban development in general. Research (Van den Berg and others, 1982; Hall and Hay, 1980; Van den Berg, 1987) has revealed that European towns and cities tend to develop through several stages of urban development. Fundamental changes in the economy, technology, demography and politics are reshaping the environment for towns and cities in Europe. The phenomenon of globalisation and the rise of the information era have induced and intensified competition among towns and cities on the regional, national and sometimes international scales. The mega trends have also effected the development of so-called *polycentric urban regions*. What are the main features of these developments and what are the implications for urban management?

The Information Era

By the 1980s it had already become clear that the rise of what was then called the 'information sector' would have a great impact on the functioning of cities and regions. Castells (1989; 1996) concludes that we have entered the informational age or era. The implications of the information era both in general and for cities can be analysed in different ways. The information era symbolises a whole set of changes that induce a transition to a new stage of urban development in Europe. What is the implication of the informational age for the functioning of cities? What trends will shape the urban economy in this era? The transition process is unfolding at full speed, changing the environment of cities rapidly.

One of the driving forces of the informational age is the rapid and continuous development of information and telecommunications technology (ICT), which in turn is bringing about a fundamental change of society. The wide acceptance of mobile telecommunication, the Internet and other new media will have a profound effect on the economy. High speed data transmission (HSDT) and videoconferencing make long-distance exchange of information possible. What are the implications of these technological developments for the use of space; that is, the spatial behaviour of people, firms and other organisations? Will the physical location remain important? Clearly, these technological developments are especially vital to economic activities that rely heavily on information and knowledge. Information and tele-communications technology has boosted the informational economy and the exchange of information, but other trends have reinforced the importance of the (exchange and production of) information as well. Information-sensitive activities prevail and will set the course for the future.

Globalisation

Technological progress seems to have 'opened up the globe' as a potential location for economic activities. An example can be found in the banking sector, which has moved its back office activities to cheap-labour countries. ICT enables such firms to follow a more cost-efficient location strategy and move activities for which physical proximity to other business units is not of the essence to cheaper locations. Globalisation is not only an economic phenomenon, for the political, social and cultural exchange across the globe has also increased tremendously.

Globalisation versus Localisation

There is another side to the 'globalisation story', however. Information exchange is not exclusively reserved to ICT channels. Personal, face-to-face exchange of information has also gained importance. The unprogrammed exchange of information in particular appears to be essential for what some (among others, Hall, 1995) have called high-touch activities (fashion, design, printing, etc.). Urban areas set the trends in high-touch sectors. A major implication is that firms are inclined to develop close relationships with their customers and to monitor new developments and trends closely. Urban areas comprise large concentrations of people and businesses. New developments very often emerge in an urban environment. Cities are the potential nerve centres of the so-called 'new economy' because, historically, cities have always been the places where information, knowledge and educational institutions such as universities and other research institutes locate.

The Rise of Polycentric Regions

Another feature of the information era is the rise of polycentric urban regions. The concept of the city is being recharged physically, functionally and also emotionally. Suburban communities have developed from mere dormitory towns to cities in their own right, with their own citizenship, employment and services. These communities have become new economic centres with their own labour market stretching far beyond the limits of the original agglomeration. The monocentric town is developing through the suburbanisation of commercial activities into a polycentric city region (Hall, 1995). The extreme examples are the 'edge cities' in the United States where the concentration of jobs and facilities outweighs the residential function.

In the polycentric region, the core city is the trademark or flagship of an urban network, often consisting of several commercial centres, with the city centre as one of the pivots. The relations among the centres in such polycentric regions are both competitive and complementary.

Competition and Attractiveness

The mega trends described above bear the stamp of urban competition. Increasingly, cities and towns behave in a logic of competition in a highly dynamic and complex environment (Bramezza, 1996). In such a competitive environment the old 'certainties' no longer exist: although the geographical situation is still relevant – cities in the centre have an easier time than cities in the periphery – it is no longer as dominant as it used to be. Qualitative location factors have come to be very important. This provokes cities to invest in their own attractiveness. But it is not enough for a town to be more attractive than its competitors as a location for enterprise. The inhabitants, the businesses, investors and visitors determine whether a city is attractive or not. These (potential) customers of cities put high demands on the quality of the business, living and visiting environment. Factors such as the level of the cultural services and access to knowledge are prominent in such an environment.

City Marketing

The developments presented above have major consequences for urban management. Urban management makes for better communication between local government(s) as suppliers of a whole range of services, and market parties as demanders for these services. To stand up to heavy competition and to satisfy the needs of the city's customers, marketing can be an important instrument in urban management; by adopting marketing principles, the municipal organisation may become more customer-oriented, ready to give service to, and mind the interests of, the city's customers.

Marketing is Instrumental to Urban Management

To entrepreneurs marketing is a useful instrument; looking at their business and their propositions through the 'eyes of the customer' helps them to reach the company's objectives (profit, market share, etc.). For urban management, we need to emphasise that marketing is instrumental to the general mission of urban management; that is, to accomplish sustainable urban development. It is important to realise that one cannot with impunity promote prosperity in terms of income and employment without giving heed to the quality of the urban living environment and living conditions of underprivileged groups in the community. But that does not exclude the use of marketing principles to attain those objectives.

Van den Berg, Klaassen and Van der Meer (1990) describe urban marketing as the set of activities intended to optimise the tuning of supply of urban functions to the demand for them from inhabitants, companies, tourists and other visitors. City marketing is both a managerial principle and a toolbox with applicable insights and techniques.

City marketing starts with the customers: the city's target groups. To the inhabitants, the city is a place to live, work and relax in, and a supplier of a wide range of facilities such as education and health care; to commuters the city is a work place; to companies it is a place to locate, do business and recruit employees; to tourists and other visitors it offers a combination of culture, education and entertainment. In other words, cities supply different functions to their variety of customers. This supply of urban functions does not stop at administrative boundaries; each function has its own functional urban region.

How do cities satisfy the needs of these target groups? In other words: what do we have in mind as the city's products? In fact there are many and manifold urban products eligible for marketing. It could be office space, harbour facilities, industrial estate or a shopping centre, but it could also be a museum, an arts festival and, of course, sports accommodation and sports events. One could draw up a long list of urban products. In general, urban products are characterised by longevity and lack of flexibility. Moreover, these products can rarely be isolated from their environment. With respect to urban tourism, Van den Berg, Van der Borg and Van der Meer (1995) make a distinction between the primary tourist product (tourist attractions) and the secondary products (for instance, accessibility).

The fact is that a city provides a 'line of products' that are difficult to isolate completely from their environment and are, moreover, highly inter-dependent. Although the city as such is not a clearly defined product, the various target groups base their decisions to locate in or visit a city on their own conception of a city; the city is then a 'brand name', so to speak.

It is difficult to say whether a city itself is a product, but the customers' associations create the city's image. There is a clear parallel with the notion of *brand marketing* that is, for instance, common for the global players in the consumer goods markets. 'Brands exists in the stakeholders' heads and hearts, not just on the sides of packages' (Duncan and Moriarty, 1997, p. 9). This notion is closely related to identity and image-building. As the spatial scope of functional urban regions widens, the relevant spatial scale is no longer only the central city but the entire agglomeration.

Organising Capacity

Kotler and others (1993) assert that increasingly the authority and responsibility for marketing and public control are delegated to private organisations, mentioning also the possibility of a community-development corporation, as a partnership between business, government, foundations and local organisations. Increasingly, metropolitan regions need to organise themselves better to improve their competitive position. City marketing makes high demands on what we call organising capacity, which is the ability to enlist all actors involved, and with their help generage new ideas and develop and implement a policy designed to respond to fundamental development (Van den Berg, Braun and Van der Meer, 1997). The performance of 'organising capacity' depends on vision and strategy, public-private networks, leadership, political and societal support, and spatial-economic conditions, as depicted in Figure 1.1.

By formulating a vision and a strategy, cities choose a direction in which they want to develop. The desired profile of a city implies strategic choices for specific sectors or clusters. Political and societal support and cooperation among public and private actors are needed to create the desired profile. Spatial-economic conditions, such as a common opportunity or threat, can stimulate the cooperativeness of the various actors.

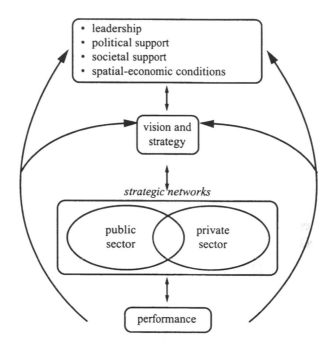

Figure 1.1 The elements of organising capacity

Source: *Metropolitan Organising Capacity; Experiences with Organising Major Projects in European Cities* (Van den Berg, Braun and Van der Meer, 1997).

The organisational complexity of city marketing depends on the definition of the product. We can distinguish three product levels: facility, cluster and city. On the first level, city marketing concerns one urban product, such as a museum, a sports stadium or a shopping centre; the differences in general marketing are not great. On the cluster level, city marketing refers to a spatial concentration of (urban) products that have a functional relation in the perception of the users; examples are clusters in the sphere of culture, sports, entertainment and harbour-related activities. In the third approach, the city itself is regarded as the brand of a range of urban products.

The three product levels can be characterised by an increasing complexity and a growing need for public-private partnerships. On the first level the product is relatively easy to define. On the second level, though, it is not always clear which urban products form a cluster in the eye of the users. The third product level brings many questions: what elements determine a city's image and how can that be related to other product levels? This implies a large number of producers, which increases the necessity of organising capacity.

The Social and Economic Value of Sports

To understand the relation between sports and city marketing, we first need to discuss the development of the social and economic value of sports. Sport was 'invented' in England in the second part of the ninteenth century. Of course, competitions – in which men could test their strength, competence and physical talents – were already organised in other civilisations but the skills that were put to the proof in such events were required for and admired in other activities, such as working, hunting or waging war. As an autonomous activity with generally approved rules, internal hierarchies, extensive organisational structures and a system of general competitions, sport is a recent phenomenon. It developed in the melting pot of private boarding schools, the exclusive education system of Great Britain. The characteristics of every sport, such as the demarcation of the playing field, the number of players, the rules and the various scoring methods, have all been developed and defined inside the walls of these boarding schools. Sports was supposed to embody the moral guidelines young men needed to play the game of life (Faure, 1994).

Since the invention of sport as an autonomous activity, its significance has increased enormously. Today, sport is an integrated part of society. For one thing, it is widely accepted that sport is good for the health, well-being and development opportunities of people. It is a positive way to spend the increasing amount of leisure time. The supply of sports facilities possibly diminishes the chance of negative behaviour in leisure time (such as vandalism) (Serail, 1993). For another, the active practice as well as the passive 'consumption' of sports contribute to social cohesion and integration. Sports provide a meeting place to contact other people (in the fight against social exclusion). Third, top-class sports (events) have developed into an international phenomenon. In 1896 the first modern Olympic Games were organised, with a limited number of participating nations. Today almost every country in the world sends athletes to this mega-event. The media (radio, television and the Internet) broadcast the results all over the world. Worldwide, television viewing figures are dominated by sports events. The matches played during the 1996 European Football Championships in England, for instance, were watched in more than 190 countries around the world by a global cumulative television audience of 6.7 billion, while some 445 million tuned in to watch the final match. Fourth, sports are big business. Internationalisation and growing media attention for sports has resulted in commercialisation and professionalisation. Since 1990, worldwide expenses on sports sponsoring have almost tripled to US$ 21,000 billion, as depicted in Figure 1.2 (Flicke, 1999). Sports and economic development are increasingly interrelated. Sports and related activities account for 2.5 per cent of world trade (European Commission, 1991).

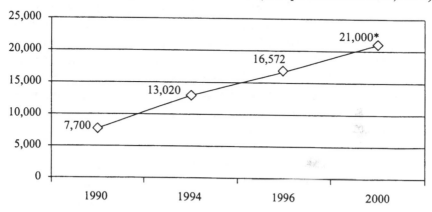

Figure 1.2   Expenses on sports sponsoring (worldwide) in millions of US$

*   Estimate

Source: *Milliardenmarkt Sport Business* (Flicke, 1999).

Towards a Research Framework for Sports and City Marketing

Increasingly, the business community sees sports, and top-class sports in particular, as an opportunity to realise their objectives by sponsoring events, clubs and sportspeople, and co-financing new sports accommodations. However, sports can also be important in the context of cities. After all, internationalisation, commercialisation and the growing media attention have considerably increased the *marketing power of sports*. So, what is the relation between sports and city marketing? In this section we will review three aspects of the interface between sports and city marketing. First, we consider the potential synergies between sports and five key elements of city marketing: the objectives of urban policy; internal and external city marketing; target groups; products; and brand marketing. A second aspect is the potential risks and problems that could undermine the synergies between sports and city marketing. The third aspect is the organising capacity to develop and monitor an integral sports and city marketing policy.

Synergies between Sports and City Marketing

First, city marketing is instrumental in the overall objective of urban management: sustainable urban development. In the competitive urban environment, a sustainable city is very much an attractive city for its inhabitants, companies, investors and visitors. In today's society, sports are an integral part of the package of amenities that make a city an attractive place to be, along with culture and the arts. In other words: an attractive, competitive city cannot do without a basic supply of sports and sports facilities. Furthermore, sports can help to revitalise deprived parts of the town with high concentrations of social problems, such as unemployment and crime (sports can reverse the negative spiral); moreover, integration of different cultures can be stimulated by sports. Second, sports as an integral part of an attractive city is first of all important for the local community (internal city marketing). At the same time, sports can be instrumental for external city marketing, serving as a vehicle for presenting the city to potential customers.

The third connection between sports and city marketing is *the urban sports product*. The sports product of a city is an element of the total urban product. It consists of all facilities and services that allow people to practise or watch sports, such as accommodation (stadiums, halls, fields), clubs and events. These facilities cannnot be isolated from other urban (sports) products.

Clubs can be very valuable to the image of a city. In Europe this applies in particular to football clubs, and in the United States to baseball, hockey, basketball and American football clubs. Professionalisation has increased the financial power of clubs, who get their income from ticket sales, merchandising, sponsors and TV rights. Some sports clubs have turned into real businesses, including a stock exchange quotation. Despite the financial volatility of the stocks, successful sports club is a valuable asset, because it is very difficult to create a new one. Some American cities have even bought clubs from other cities to create a sports image. Successful clubs increasingly generate direct and indirect employment. At the same time these clubs potentially contribute to the social structure, for their success results in a proud population (civic pride). In the social context, small clubs are at least as important as professional clubs, since they offer people the opportunity to practise sports themselves.

Cities increasingly become aware that *events* in general, and sports events in particular, are perfect opportunities for city marketing. Once-only events bring relevant actors together, which offers unique possibilities of realising complex and large-scale projects. The classic example of a city that used an event as catalyst for socioeconomic revitalisation is the city of Barcelona, which took optimum advantage of the Olympic Games. Annual events (such as a marathon or a tennis tournament) contribute to a city's sports image. Some examples are Wimbledon, Roland Garros and the Rotterdam marathon. Furthermore, events generate employment and (positive or negative) media attention.

Accommodations, such as sports halls, gyms, swimming pools and football fields, are essential elements of the urban sports product. In relation to city marketing, they raise the service level of a city and its image as a sports city, which (potentially) attracts people and businesses. Furthermore, large-scale sports accommodations can act as a catalyst for the social and economic revitalisation of their environment. They attract new activities – such as hotels, restaurants and shops – and possibly result in the creation of a cluster of entertainment and leisure-related activities. Van den Berg and Braun (1998) investigated the possible role of so-called domes in urban revitalisation and concluded that an integral vision is needed to actually use a multifunctional stadium as a motor for the urban economy.

Fourth, these different elements of the sports product are developed for various target groups. The distinction between internal and external city marketing is just one step; it is important to identify the more specific target groups that are the customers of the various urban sports products. What is the profile of the urban customers for these products?

Fifth, the city's sports product can contribute to the image and identity of a city. So, the city management might not just support a sports event on behalf of the participants and visitors, but also see it as part of its 'brand marketing' strategy. A very good example of a community that has followed a clear brand-marketing strategy is Indianapolis. Many considered Indianapolis dull and boring; someone even called it 'a cemetery where the lights are put on once a year for the famous racing event, the Indy 500'. In the late 1970s politicians, business leaders and other stakeholders in the community identified sports as a vehicle to improve the image and attraction of the city and the downtown in particular. They reckoned sports to be one of the few things suitable to create an image around. Their strategy centred around sporting events with a clear build-up from smaller to medium-sized and some big events. The city discovered amateur sports as a market niche through which the city could develop its sport profile and promote it as a *sports place*. Through the staging of events like the Pan Am Games, the World Gymnastics Championships and the Final Four (basketball) the city communicated that it was an attractive place to live in and to visit. These events, together with the two major league professional franchises, have portrayed the city as a sports city and today many refer to Indianapolis as the 'Amateur Sports Capital'.

Risks and Potential Problems with Sports and City Marketing

According to Oldenboom (1999), the socioeconomic value of sports (events) is built up by tangible (financial) and intangible effects. The first group generates extra expenditure and employment in the metropolitan region. Intangible effects can be subdivided into general intangible effects (externalities) and specific intangible effects. General intangible effects are nonexclusive: everybody takes (dis)advantage of these effects. Most of the potential synergies mentioned above could be ranked in the category 'intangible'.

In discussing the contribution of sports in relation to city marketing, we also need to address the possible risks and drawbacks of investment in sports. Investment in sports events, clubs and accommodations often comes out of governmental budgets. Cities that organise major events should take into account that there is the risk of (large) deficits which might be harmful to urban development. A well-known example of a city that experienced the negative value of (sports) events is the Canadian city of Montreal, which suffered under the deficits resulting from the Olympic Games in 1976. Another point is that sports venues use scarce public space and possible resources. The construction of a new stadium might be at the expense of space for apartments, offices or a green area. Finally, the large crowds drawn by sports events that might cause disturbances in the city, such as traffic jams, plundering and confrontations between 'supporters'. The phenomenon of hooliganism as a result of sports events is associated mainly with football.

Organising Capacity for a Sports and City Marketing Policy

The use of sports in city marketing implies investment in the urban sports product (product development) and the use of sports for the promotion of the city. Ideally, that strategic choice is related to a general vision and strategy with regard to the city, based on internal strengths and weaknesses and external opportunities and threats. Consequently, cities with a comprehensive and high-quality urban sports product are well placed to use that strength for their city marketing strategy. The optimum capitalisation of the potentialities of sports and city marketing makes high demands on the city's organising capacity. What parties are involved? What are their ambitions? Who will assume the role of pioneer? Is there enough political and societal support? Does the sports and city marketing strategy fit the vision and strategy of the region?

In Sum, the Research Approach

In the previous pages we have introduced the subject of sports and city marketing in the wider context of urban development. We have taken a closer look at the role of sports in city marketing and that resulted in a research framework that we described above. The framework incorporates positive and negative interactions between sports and city marketing as well as the organisational challenge. The elements that are depicted in Figure 1.3 are the guidelines for the five case study chapters that follow hereafter. The research approach in the five case studies consists of desk research (policy documents, other research documents, bid books etc.), site visits and in-depth interviews with key figures in the city both from the public and private sector, in order to confront the ideas in this chapter with the practice in five European cities.

Positive interactions
1 The city's sports supply:
      regarding sports as part of an
      attractive city
2 Product development
      creating a good sports product
3 Brand marketing:
      promoting and imaging the city
4 Differentiated marketing:
      internal and external marketing

Negative interactions
1 Financial problems:
      budget deficits and lack of resources
      for other urban products and services
2 Spatial problems:
      lack of space for other urban products
      and services, accessibility problems
3 Safety problems:
      crowd-control, hooliganism
4 Lack of local support:
      increasing entry barriers for local
      community

Organising capacity
1 Strategic networking:
      what parties are involved?
2 Leadership:
      who assumes the role of pioneer?
3 Vision and strategy:
      do sports and city marketing policies
      fit the vision and strategy of the
      region?
4 Support:
      is there enough political and societal
      support?

Figure 1.3   Research framework for sports and city marketing

Chapter Two

Barcelona

Introduction

Undoubtedly, Barcelona belongs to the premier league of European sports cities. The second city of Spain owes this status to its successful clubs and its experience in organising major sporting events. The sports image of the city is largely determined by Football Club Barcelona, which has more members than any other football club in the world and the second highest budget after Manchester United. The four professional sports departments of the club (football, basketball, handball and roller hockey) all play at the top European level. Furthermore, at the time of writing Barcelona's second football club – Espanyol – also plays in the Primera Division (the Spanish premier league).

Another important element of Barcelona's sports image is the fact that the city hosted the Olympic Summer Games – the biggest event in the world – in 1992. This event has resulted not only in a high-quality sports infrastructure, but also in a more attractive city. Therefore, Barcelona is probably one of the classic examples of how to use (sporting) events for city marketing. Eight years after the Olympics, the city is still very active in that field, as witness its efforts to use a brand new event (the Universal Forum of Cultures in 2004) to revitalise an old industrial area.

One of the most interesting developments in the field of sports related to the city marketing of Barcelona concerns an initiative by Football Club Barcelona. In 1999, this club publicised a plan to create a sports-related entertainment park in the area of the Nou Camp stadium. On the one hand, the construction of such a theme park may help to promote the city. On the other hand, it could have some negative effects on the city's attractiveness.

This chapter discusses the various topics mentioned above in more detail. After a brief profile of the city, the legacy of the Olympic Games is reviewed. The next section deals with FC Barcelona and its plans for a theme park. Next, we will concentrate on Barcelona's city marketing strategy and the role of sports in this strategy. Some conclusions will complete this chapter.

Profile

Barcelona is the capital of Catalonia, an autonomous region in Spain with six million inhabitants. More than 1.5 million citizens live within the city limits, while its metropolitan region has a population of more than 4.2 million. Until 1714 Catalonia was an independent state. Although the region now belongs to Spain, it still has its own culture and language. After the fall of the Franco regime, Catalonia was given the status of autonomous region, with a government (the Generalitat) and parliament of its own. Moreover, in 1983 the Catalan government established the Catalan tongue as the second official language, which made Catalonia formally bilingual. At present, Catalan is the spoken language for most citizens. In spite of their common cultural background, the Generalitat (Right Wing in 1998) and the Barcelona City Council (Left Wing in 1998) have rather different ideas.

The city of Barcelona is divided into 10 districts. The historical centre of the city is the district Ciutat Vella ('old city'), which comprises a mediaeval street pattern with very narrow streets and lanes and blocks of houses mostly of six storeys. This neighbourhood is separated from the seafront by a busy thoroughfare. This seafront consists of (among other things) marinas and promenades with various restaurants. Between 1870 and 1936, the city centre was been extended by the construction of the Eixample district (Catalan for 'extension'), designed by Cerdà, a famous architect. Eixample has a very characteristic grid system, which is made up of square blocks, each with a length of 113 metres, with cut-off outer corners, producing more or less regular quadrangles. The third district worth mentioning is the Sants Montjuïc district, where the Olympic Stadium is situated. The statistics, especially those about home-to-work traffic, show that Barcelona is steadily expanding its influence. A vast area has become intra-dependent, so that a polycentric region has emerged: a network within which certain cores and suburban outskirts are gaining weight (Nel.lo, 1998).

The Legacy of the Olympic Games

The 1992 Olympic Games (OG) marked a turning point in the history of Barcelona. The Catalan city can be regarded as a classic example of how to use a major sports event as a stimulus for product development and city promotion. The games made a considerable contribution to the metamorphosis of an industrial harbour city into an attractive and service-oriented

Mediterranean city. The city council used the event as a lever towards political consensus for a large-scale urban renewal programme, which was a crucial part of the bid-book. After Barcelona's assignment in 1987, a strategic plan (Barcelona 2000) was drawn up, based on the assumption that the investments related to the Games should primarily aim at long-term effects. The basic idea was to use the event to put into practice a vision that already existed: to consolidate Barcelona as an enterprising European metropolis (Ajuntament de Barcelona, 1990) by making the city more attractive.

The investments (product development) can be divided into two categories: general investments and sports-related investments. Examples of the first category are the modernisation and expansion of the airport, the construction of a ring road, the revitalisation of the city centre and the spectacular waterfront development (which effectively turned the city's face around to the sea). The second category relates to the renovation of existing facilities and the construction of new ones.

The locations of the Olympic venues were chosen to fit the objectives of the urban planners with a view to a better balance between the (too) fast-growing southwest and the somewhat lagging northeast, more emphasis on the prospects of the urban periphery (rather than too much pressure on the central districts), improving the building on Montjuïc Hill and opening up the town towards the seafront. The investments in sports accommodations took up only 9 per cent of the total investments related to the Olympic Games. They include the renovation of the Olympic Stadium and the construction of 15 indoor venues.

Multifunctional Venues

Barcelona deliberately decided to develop *multifunctional* venues to guarantee their future usefulness. Consequently, most venues offer space for various activities in the sphere of sports, culture and commerce. A good example is the Olympic Stadium, which is presented as 'the place where everything is possible'. Owing to a combination of football (the home matches of Barcelona's second club *Espanyol*) and American football (Barcelona Dragons), concerts and commercial activities, this stadium is operating profitably.

The concept of sustainable accommodations is opposite to that of temporary accommodations (which have been used for the Olympic Games in Atlanta, for example). However, a compromise between the two concepts is also an option: some Olympic venues in Barcelona were equipped with temporary stands. In general, the Olympic venues had to meet the demands of the IOC (such as a minimum capacity) and those of (future) citizens. Nevertheless, the question 'what to do with a venue after the event?' was not answered in advance for all facilities. Afterwards, many accommodations had to be adjusted to meet the demands of the new users. After the Games were closed, however, financial support for such investments appeared very hard to get, since public money was very scarce in that period of global recession. Moreover, some venues appeared less useful than expected. For instance, the *monofunctional* basketball stadium stood idle after the Games for quite a while,[1] which illustrates the importance of multifunctionality. The occupancy rate of some other facilities with only one large unit fell short, which supports the plea for multi-unit facilities[2] to cope with space consumers' different preferences.

One of the legacies of the Olympic Games is a generous supply of sports facilities. After the event, most facilities were privatised and handed over to public companies (by 'public tenders'). The owners of the facilities have to fulfil several conditions, such as accessibility for everybody, fair prices and special programmes for the elderly, young people, immigrants and low-income groups. The occupancy rate of the majority of the venues is extremely high. Sports clubs (and fitness clubs in particular) are very popular and many people are on waiting lists for membership that will give them the right to a relatively cheap season ticket. Membership is stimulated by a rise in single ticket prices.

Tourism

Another legacy of the Olympic Games relates to tourism. Before 1990, the city was visited only by people with business purposes and by tourists who were actually staying outside the city at one of the Costas. As depicted in Table 2.1, the number of tourists increased from approximately 1.7 million in 1990 to almost 3 million in 1998.[3] In the same period, the number of overnight stays almost doubled from 3.8 million to 7.4 million, while the share of leisure tourists grew steadily (see Table 2.2), the number of meetings and congresses tripled from 373 to 1,002 and the number of hotels increased from 118 to 166.

Table 2.1   Number of overnights and tourists (in thousands)

	1990	1992	1994	1995	1996	1997	1998
Overnights	3,796	4,333	4,705	5,675	6,341	6,965	7,400
Tourists	1,733	1,874	2,664	3,090	3,062	2,823	2,969

Source: *Tourism Statistics 1998* (Turisme de Barcelona, 1998).

Table 2.2   Tourists by purpose of visit (%)

	1990	1992	1994	1995	1996	1997	1998
Vacation	22.7	30.8	31.3	35.3	36.4	42.3	51.8
Business	53.8	53.7	47.4	42.1	43.2	39.5	35.0
Congresses	4.5	5.3	5.4	6.6	6.3	5.6	8.6
Fairs	10.8	3.0	5.1	7.2	5.4	5.1	3.7
Family visits	4.5	4.7	3.9	1.7	2.2	1.9	0.5
Others	3.7	2.5	6.9	7.1	6.5	5.6	0.4

Source: *Tourism Statistics 1998* (Turisme de Barcelona, 1998).

The Olympic Games put Barcelona on the map as a tourist destination in two ways. Firstly, the urban renewal project related to the event has improved the quality and the internal accessibility of the tourist product, while the extension of the airport has enhanced its external accessibility. Secondly, the event itself generated a lot of publicity: potential free advertising for the city. The city was on the front page of more than 15,000 newspapers around the world, with a total estimated circulation of 500 million copies. Twelve thousand accredited journalists covered the Games, which was not only more than for any previous Olympic event but also a record in the world of news coverage in general (De Guevara, Cóller and Romani, 1995).

Before, during and after the Games, Barcelona received much positive media attention, such as stories about the unique urban renewal project, the cultural heritage of the city and the economic strengths of the region. Of course, the media reported on negative elements as well. One year before the Games, one of the most important news items for the international press was the security of the Games and the risk of terrorist actions. However, the positive development of political conflicts that might have influenced attempts at violent action, and the neutralisation of terrorist groups thanks to a thorough police clampdown, meant that, in the end, security became one of the items which least interested the international press (De Guevara, Cóller and Romani, 1995). The experiences of Barcelona illustrate the importance of avoiding the negative aspects of a sports event in order to optimise its city marketing benefits. Eight years after the Games, the Olympic spirit is still alive in Barcelona.

Barça – More than just a Sports Club

Obviously, the Olympic Games proved to be a unique opportunity to create a new city and a new image. However, despite its long-term benefits, it is a one-off event. An example of a recurring event in Barcelona is a Football Club Barcelona match. This club was founded on 29 November 1899. In the course of 100 years, political developments have made a great impact on the club. During Franco's dictatorship, FC Barcelona turned into a symbol of opposition to the central government: a symbol to underline the differences between Catalonia and the rest of Spain. The club's stadium (Nou Camp) became a place where nationalist feelings could be expressed by flying the FC Barcelona flag, which in a way replaced the flag of Catalonia, forbidden at the time. As a consequence, the club welcomed members with different political views and from various social classes. From the 1940s onwards, the number of members increased from 25,000 to 100,000 in 1982.

In the course of time, the club has put down roots in the society and become a symbol for independence. Although Catalonia is now an autonomous region in a democratic country, many people still fear or distrust the national government, represented in the eyes of many FC Barcelona fans by the football club Real Madrid. Most inhabitants of Catalonia look upon Barca (the club's nickname) as their 'national team' and a sign of hope. The number of members reached a peak in 1986 with 109,000 *socios* (Spanish for 'members'). Since then, it has never dipped below the 100,000 mark. Currently, about 104,000 *socios* have joined the club. Since 1978 the number of supporters' clubs has increased from 96 to 1,300. The network begins in Catalonia and extends throughout Spain to the rest of the world. In 1999 the club celebrated its 100-year anniversary. The celebrations included several cultural and sporting events, such as a sports medicine seminar, a concert, an exhibition and several matches (in a range of disciplines). The club participated actively in some of the city's main events: the Corte Ingles races and the Gymkhana events at La Merce festival.

FC Barcelona's football team, which is largely responsible for the image of the club, is one of the most successful teams in the world. Barça's trophy collection includes one European Cup, four Cup Winners' Cups, four UEFA cups, 24 national cups (Copa del Rey) and 16 Spanish League Cups. Since August 1999, the FC Barcelona premier league games have been broadcast by Canal Barça, on the basis of a pay-per-view system. Furthermore, FC Barcelona's TV station offers subscribers the latest club news, interviews, historical matches and coverage of basketball, handball, roller hockey and lower division football games. Barcelona's basketball team belongs to the European top teams, with five performances (and losses) in the European League final (comparable to the Champions League). In the 1980s and 1990s, the team won several national titles. The Barcelona handball team is undoubtedly one of the best in the world and boasts an impressive trophy cabinet containing several European and national awards. Finally, the roller hockey division has more national and European titles to its name than any other section of the club.

Organisational Structure

Unlike many other Spanish clubs and European top clubs, FC Barcelona is not a company. In 1993 – when many clubs were hit by a financial crisis – a Spanish law obliged most clubs to turn themselves into a company in order to get capital and to make the club a healthy organisation. Four clubs, including Barça, held an exceptional position, because of their positive financial situation at the time. To ensure that FC Barcelona could remain a club and would not have to become a limited company, a Foundation was created in 1994.

The main task of this Foundation is to cover the costs of all non-profit activities of the club, including the activities of the eight amateur divisions (athletics, field hockey, rugby, baseball, volleyball, ice hockey, figure skating and indoor football) and the youth divisions of the four professional sports divisions (football, basketball, handball and roller hockey). The Foundation is also responsible for organising the annual congress of FC Barcelona supporters' clubs as well as financing non-profit cultural activities within the club. Furthermore, the Foundation takes care of promoting sports among youngsters (the number of youth players is approximately 2,000), financing the education of young players and supplying accommodation to children whose parents live outside the Barcelona region (they stay in the La Masia residence). Special apartments in the city have been arranged for young adults and the Foundation also keeps in contact with schools and universities in the city to offer them a good education. In 1999, the Foundation received an award from the King of Spain for its contribution to amateur sports in Spain.

The Foundation's budget is mainly financed by the yearly contributions of private companies. Membership can take several forms. Honorary Members contribute €60–90 million,[4] while Associate Members pay more than €21,000. In return, they receive a comprehensive hospitality package, including seats, parking spaces, tickets and access to VIP spaces. At the time of writing, all membership cards available in these two categories are allocated. However, it is still possible to become a Patron Member by contributing more than €6,000 a year, which gives the right to one seat and a parking place. Corporate membership is obviously very popular. Firms from Catalonia, Spain and the rest of the world have decided to become members, usually not for reasons of publicity, but to keep in touch with clients by making use of the hospitality packages. The Foundation also accepts members who wish to make personal contributions: a one-off fee in exchange for a plaque with the member's name engraved upon it at the club's stadium. Up to the time of writing, more than 23,000 people have supported the club in this way.

Most members of FC Barcelona want to maintain the present organisational structure, which gives them say in the club's policy by electing the members of the board.[5] In that respect, they feel they 'own' the club, unlike a legal company, which is owned by the shareholders. At the same time, the club wants to safeguard its position in the national and European competitions. To attain both objectives (maintaining a club and being successful) will be difficult, since many other European clubs turn themselves into legal companies, go to the stock exchange, receive a lot of extra capital and buy the best players.

Most of Barça's income is generated by television rights and merchandising. Currently, the club is investigating possibilities of making more use of sources of income that other European clubs use in abundance. The club's first proposal was to introduce advertising on the players' shirts, but most members protested against that idea: they wanted to keep the team shirt 'clean'. Next, the devaluation of the season tickets, which now entitle the holders to visit all matches including Champions League games (which is quite extraordinary compared to other clubs), was considered. Thirdly, skyboxes (the very first in Nou Camp!) are now under construction to meet the demand for such facilities. Finally, Project Barça 2000 (see below) also fits the club's strategy to create more sources of income.

Project Barça 2000

FC Barcelona is the owner of an area on the southern edge of the city centre. Nou Camp, which is located in this area, is the largest stadium in Europe, with a capacity of 98,000. El Nou Estadi del Futbol Club Barcelona was opened in 1957 and soon began to go by the popular Catalan nickname of 'Nou Camp' (new stadium). The UEFA (the European Football Association) has awarded the stadium the maximum number of five stars. Therefore, Nou Camp is allowed to host Champions League finals, as happened in 1999. The stadium receives many tourists for a visit to the Club Museum located in the stadium grandstand. It is the second most visited museum in Catalonia after the Picasso museum.

In 1999, the club presented a plan to turn this area into Barcelona FC Park. The area can be divided into two parts: the stadium area (around Nou

Camp) and the Mini-Stadium area (around the small stadium for amateur sports). One of the main elements of the plan is an open urban park that covers both areas, as well as a 50 metre wide bridge (to be constructed) across the street dividing the two areas. These public spaces cover an area of 55,000 m^2 (see Table 2.3) (FC Barcelona, 1999). According to the plan, the basic sports facilities (Nou Camp, the Mini-Stadium and Palau Blaugrana) would be retained, while the other sports facilities (mainly for training purposes) would be moved to a new training complex, 4 km south of Nou Camp.

Table 2.3   Activities planned in the Barcelona FC Park

Multipurpose service zone	Cultural zone	The Boulevard	Entertainment zone
Hotel	FC Barcelona museum	Shops	Cinema
Club offices	IOC museum	Restaurant	Sports World Pavilion
Members Centre	Sports museum		American Sports Pavilion
Members Hall	Centenary Pavilion		Cinema World Pavilion
Sponsor Area	Barça Megastore		
VIP zone			
TV-studio			
Convention rooms			
Fitness centre			

Source: *Parc del Barça; The Project* (FC Barcelona, 1999).

The theme park, as proposed by FC Barcelona, is a combination of culture, leisure and entertainment related to sports and more specifically to football and FC Barcelona. The project had two main objectives: on the one hand it aimed to create a central area in which the club and its urban surroundings blend together to offer visitors a high-quality environment. On the other hand, it aimed to open prospects for further development in the leisure sector and new services for people. Apart from the sports facilities, the area would be divided into four zones. Table 2.3 gives an overview of the various activities planned in each zone, including museums, restaurants, shops, thematic pavilions with exhibitions and virtual reality applications and a multiplex cinema. The space would be allocated to the different components as depicted in Table 2.4.

Table 2.4   Allocation of space in the Barcelona FC Park (m²)

Public spaces	55,000
Cultural spaces	4,000
Recreational areas and pavilions	27,000
Multicinemas	9,000
Restaurants and retail space	20,000
Multipurpose service buildings	30,000

Source: *Parc del Barça; The Project* (FC Barcelona, 1999).

According to the plan, the number of visitors to the Barcelona area would increase from 2.5 million to more than 7.5 million. FC Barcelona (1999) expects that half of the visitors will come by car, as depicted in Table 2.5. The plan comprised the construction of 4,000 parking places and advised the enhancement of accessibility by public transport, for instance a shuttle service by a light monorail to connect the bus stops and underground stations on the Diagonal with the FC Barcelona Park. This shuttle should encourage the use of public transport, as in the existing situation people have to walk for more than 10 minutes from either of the two nearest underground stations.

Table 2.5   Means of transport used by visitors

Car	50%
Bus	5%
Underground	20%
Motorbike	4%
Taxi	1%
Pedestrians	3%
Private coach	17%

Source: *Parc del Barça; The Project* (FC Barcelona, 1999).

Except for the buildings on the outer perimeter and the sports facilities, there would be no significant building in the Park rising above the esplanade that extends to and covers the whole area. The execution of the project would be handed over to an investor. FC Barcelona would remain the owner of the land and rent the buildings using long-term concessions. The removal of the training complex is an independent project, financed by FC Barcelona.

To realise the plans for the Barcelona Park, permission from the municipality was needed, because the function of the area was to be changed (from sports only to entertainment, culture, leisure and sports). The approval of the Catalan government (the Generalitat) was not needed in this case, because the project did not change the general zoning plan. The municipality is authorised to make special zoning plans that fit within the general zoning plan. Because the plans were related to such a special zoning plan, they had, however, to be made public.

The publication of the plan generated a lot of criticism from politicians, urban planners, architects, a neighbourhood association and the media. Some of the opponents to the plan accused the club (and its president in particular) of land speculation, which has been denied by the club. Furthermore, the urban planners, architects and the neighbourhood association argued that FC Barcelona's plan does not consider the effects on the plan area's surroundings. They feared that the more than 5 million extra visitors expected by FC Barcelona (1999) would diminish the internal and external accessibility. Accessibility in this part of Barcelona is very vulnerable, because of the high concentration of residents and public facilities on the one hand and the already-congested road infrastructure on the other. The crossing of Barcelona's main access road from the south with the ring road (Rondas) and the Diagonal, and the presence of Nou Camp, a hospital, a technical university and a technology park generate a lot of traffic.

Although the plan included measures to improve access by public transport, it also stimulated private transport by offering 4,000 parking places. According to opponents of the plan, the theme park would have a negative impact on the accessibility of the entire region, because of its location near the main access roads to the city. In their opinion, the construction of the park would actually imply a change in the general zoning plan (because of the accessibility effects) for which permission from the Generalitat would have been needed.

The *legal* opposition to the project was divided into two groups, the urban developers and architects and a neighbourhood association. Both groups have instituted legal proceedings against FC Barcelona. In January 2000 the city council proposed to create a new plan for the Barcelona area *and* its surroundings to solve problems (related to accessibility for instance) instead of creating new ones. Because of that intervention, the lawsuits were cancelled and the media seemed for the time being to have suspended their opposition against the plans. The project will be managed by the city of Barcelona and involve some other municipalities in the metropolitan region. A team of professionals (such as architects and economists) responsible for guiding all public and private investment projects in Barcelona will supervise the project team. The ambitious plans of FC Barcelona will be reconsidered and adapted if necessary. It is not yet clear what will remain of the original plan.

The discussion between opponents and advocates of the plan will probably continue. Presumably, FC Barcelona will try to maintain the most attractive elements of its plan to keep it commercially attractive, while opponents will argue that the number of attractions should be diminished to reduce the number of visitors. A common vision is needed to define a strategy that can be supported by both groups. Such a vision should preferably be based on common interests: firstly, the project can generate employment, attract visitors and reinforce the city's (sports) image; secondly, improved road infrastructure and public transport (such as a new underground line or a higher frequency) would benefit all actors involved; thirdly, the plan might result in a more equal distribution of visitors over the week, thus relieving the pressure on the area's accessibility.

Barcelona's City-marketing Strategy

Like many other 'second cities', Barcelona aims at a specific profile to obtain a status comparable to that of European capitals ('first cities'). First cities have a strong public sector and receive many state investments. Barcelona has succeeded in obtaining the status of Event City and in using the events as catalysts for urban development. This strategy was already in use in 1888, when an international exhibition stimulated the construction of Eixample's grid system and the industrialisation of the city. An international exhibition in 1929 resulted in the construction of the underground system and the city's westward extension. In 1952, when Franco was in power, a Catholic congress was used as the lever to get the money needed for the construction of houses for immigrants from the central government.

When Spain became a democracy in the late 1970s, Barcelona started to work out a strategy to develop the city in a positive direction. The city realised that sports in general, and the Olympic Games in particular, could play a part in that strategy (see 'The Legacy of the Olympic Games', above). A very important element in the sports image of the city is FC Barcelona (see 'Barça – More than just a Sports Club', above). Furthermore, Barcelona has a long tradition in sports, with such annual events as the Formula One Grand Prix, the Conde de Godó International Trophy (one of the nine ATP tournaments) and about 30 other events.

Additionally, Barcelona is a centre of knowledge concerning sports and sporting events in particular. The Autonomous University of Barcelona offers courses in sports-related topics and instigated the creation of the Olympic and Sports Studies Centre in 1989 for purposes of research, documentation, training and dissemination of information on Olympism and sports. This centre is governed by the Barcelona City Council, the Spanish Olympic Committee, the Autonomous University of Barcelona and the Barcelona Provincial Council. The centre has been responsible for several publications (to be found in the Olympism and Sports library), conferences, seminars and symposia in the field of sports and the Olympic Games. Since 1995, the centre has managed the International Chair of Olympism, which aims to stimulate research, education, documentation and dissemination of the Olympic movement's ideals.

Promotional Activities

Barcelona is promoted by several organisations on various spatial levels. On the municipal level, Turisme de Barcelona aims to promote the city as a tourist destination. The City Council, the Chamber of Commerce and the Foundation Barcelona Promoció founded this organisation in 1993. Since 1994, Barcelona's promotion strategy has been changed from a general to a differentiated approach. Each promotion campaign is tuned to the demands of a specific target group: a market niche. Consequently, Barcelona is presented in several ways: as a destination for theatre lovers, young people, gays, people who like classical music, etc. Moreover, Turisme de Barcelona decided to put the city on the map as a 'city trip' destination rather than a 'sun and beach' destination. After a year of negotiating and paying to be included in the brochures, tour operators were convinced of the qualities of the city and could no longer afford to exclude Barcelona.

The target groups of the city promotion change from time to time. Turisme de Barcelona is continuously on the look-out for new target groups by contacting relevant organisations in foreign countries. For instance, the organisation works together with tour operators who arrange city trips for gays. For that target group, the city presents itself as tolerant and liberal. This way of promoting the city is considered more effective and much cheaper than forms of general promotion (such as a stand at a holiday fair). Consequently, Turisme de Barcelona does not present the city as a sports city to *all* people, but only to a limited number of target groups. The organisation considers FC Barcelona one of the ambassadors of the city, but the club is hardly involved in city promotion.[6]

The target group strategy is complementary to the more general promotion strategies of the Catalan and national tourist boards. These organisations have a budget for advertisements, while Turisme de Barcelona uses its budget to invite tour operators (to show them the quality of the city) and journalists (who write believable stories about the city). The budget is financed by the municipality, the Chamber of Commerce and the resources generated by such tourist products as Barcelona Bus Turistic, Barcelona Card, Barcelona Pass, Barcelona Walking Tours, Columbus Monument and the Horse Show in Barcelona. These products try to make the city as accessible as possible by offering visitors a wide choice, thus enabling them to discover the city and enjoy their stay to the full.

The municipality wants to maintain the balance between the tourist function and other functions (such as the residential one). Such a balance is needed to keep up the image of a vibrant Mediterranean city. Although opinions about the carrying capacity (the maximum number of tourists) are much divided, some growth in the number of visitors seems to be acceptable. Investment in tourist attractions are primarily aimed at keeping the number of tourists on the same level (without these investments, their number would decline). Furthermore, there is a growing need for hotels, especially in the business segment (four and five stars), which is illustrated by the high occupancy rate (more than 80 per cent in 1998). In the course of 2000, 18 new small and medium-sized hotels were opened, supplying an extra 1,500 beds. Within three years, some new large hotels will open their doors in Barcelona.

Barcelona has a reputation for using events not only for promotion but also for urban renewal plans. The city has even *created* a cultural event to function as catalyst for a specific area. This area is located south of the Olympic Village and used to have an industrial function. It is to become a so-called 'eco park' as a model of urban sustainability. Table 2.6 gives an overview of activities planned in the area.

Table 2.6   Activities planned in the Universal Forum of Cultures area

Peace centre	Outdoor meeting place	Service areas
World diversity	Marine exhibition area	Swimming and water games area
Digital convention centre	Aquatic theatre	Treatment plant and solar energy park
Sustainable city	Amphitheatre	Solid-waste treatment area
Zoological experience	Exhibition park	Electricity park

Source: www.barcelona2004.org.

Until 2004 – when the Universal Forum of Cultures will take place – each year will have a specific theme that draws attention to that event. Promotional activities will be adjusted to this theme. Music was the theme of 2000, followed by art in 2001. The famous architect Gaudi will be celebrated in 2002. The year of sports will be 2003, because of the various sporting events to be organised, such as the World Swimming Championships, the Police and Fire Games and the World Championships of (Field) Hockey. All these events can be used to draw the world's attention to the Universal Forum of Cultures in 2004. One event, the World Swimming Championships, can be exploited to enrich Barcelona's image by emphasising some of the city's less familiar strengths such as the sea, the beach and the Olympic harbour. An interesting finding is that events can be used to make a city attractive, but cities can also be used to make sports attractive. The organiser of the championships – the World Swimming Association (FINA) – has selected Barcelona, which is considered an attractive city with a strong image, to reinforce the image of swimming. To meet the demands of both the city and the swimming association, two new swimming pools with a view of the sea (one permanent and one temporary) will be built, and the sea itself will be used for long-distance open-air swimming.

The Use of Sports for Social Objectives

The city uses sports not only for city promotion (directed at external target groups) but also for social objectives. Sports facilities are obliged to reduce their prices for elderly and young people, immigrants and minimum-income groups. The city wants to increase the accessibility of sports to all people. Every citizen has the right to participate in sports (by buying a ticket).

Ethnic minorities are stimulated to participate in sports by special football competitions (although of course they can also play in the regular competitions). Moreover, the city deliberately builds sports facilities in neighbourhoods with a weak social infrastructure and offers courses to people who would like to become referees. The City Council is considering the introduction of a discount card for low-income groups to increase the accessibility of facilities in the sphere of sports and culture. One obstacle is the high number of illegal immigrants, who will never make use of such a card. FC Barcelona is not involved in any of these social programmes. Barça footballers do not promote sports in the neighbourhood.

The involvement of the population is crucial for the success of a sports and city marketing policy. First of all, residents should be able to participate in sports, an objective that is now thwarted by the long waiting lists for clubs (swimming, cycling, running and fitness in particular are currently very popular). At the time of writing there are no concrete plans to build new sports facilities. Another threat is the danger of performing sports in public. Cyclists and runners have to leave the city to find a safe place. Secondly, people should be able to watch sporting events in the city. When Barcelona hosted the Champions League Final in 1999, it was all-but impossible to buy tickets for this match because most were reserved for VIPs. The citizens could not make the event their own but were rather alienated from it.

Conclusions

Barcelona's experiences have proved that potential synergies can be obtained from using sports in city marketing. The Olympic Games have been used as a catalyst for the creation of one of the most attractive cities in Europe. The mega-event stimulated improvements in public and private transport, investments in the quality of the living environment and the socioeconomic revitalisation of the city centre and the seaside. Obviously, Barcelona City Council looks upon sports as an indispensable element of urban attractiveness.

Another positive legacy of the Games is a comprehensive sports product, including venues and experience in organising sports events, giving the city a competitive advantage in attracting sports events. Moreover, Barcelona was able to use the event for city promotion. Thanks to the media attention, Barcelona's image has improved considerably, which has probably increased the number of visiting tourists. The sporting, economic and promotional success of Barcelona '92 is the result of an integral vision and strategy that was formulated in cooperation with all relevant actors in the Barcelona region. Clearly, the event and the construction of accommodations (an element of the sports policy) were an integral part of the overall strategy aimed at sustainable economic development.

After the Olympic Games, Barcelona entered a new stage of using sports in city marketing, characterised by a shift from a general to a more differentiated approach. Today, Barcelona uses sports as one of the elements of its city-marketing strategy, which can be subdivided into several sub-strategies for each external target group. The city has also developed policies directed at internal target groups, using sports as an instrument against social exclusion.

The construction of new accommodations has no priority, but the organisation of sports events to consolidate the status of Event City is still regarded as very important. Barcelona has used a sports event to improve its image. Now, sports events have begun to use Barcelona to improve the image of an event or a particular sport. Annual sports events in the Barcelona region also benefit from the enhanced image, because participants are very willing to visit the city.

One of the most interesting developments in Barcelona is FC Barcelona's proposal to construct a sports-related leisure park. Potentially, the park has positive and negative implications for city marketing. Such a complex will probably generate much additional tourism and consolidate the sports image of the city. Moreover, it will benefit FC Barcelona and help the club management to safeguard its position in Europe without changing its organisational structure.

Currently, the club and the citizens have a very strong relation, but that situation could change if FC Barcelona decided to convert the club and its Foundation into a company. FC Barcelona Park would be one way to earn money from new activities. Opponents of the project, however, are concerned about its potential consequences for the surrounding area. The project might adversely affect the accessibility of the FC Barcelona area, and even put pressure on the transport infrastructure of the entire urban region.

However, FC Barcelona and its opponents have shown willingness to discuss compromise solutions to arrive at plans for a park that benefits the city without creating new (spatial) problems. In this process, the leadership role has shifted from the club – which came up with an innovative and ambitious project – to the City Council, which is trying to integrate the plan into the general strategies of the city. The leading role of the city should make it possible to optimise the synergies from this project and minimise its negative effects on accessibility and the quality of the living environment. In view of the distrust against FC Barcelona's club management, the city should take the lead in creating societal and political support for the plans.

From its Olympic experiences, it can be concluded that Barcelona has a very strong organising capacity, characterised by an integral vision on urban development. Nevertheless, the local government still faces the challenge of strengthening the relationship with one of the most important actors with regard to sports and city marketing: FC Barcelona. Although this football club is an essential element of the city's sports image, it is insufficiently activated in internal and external city marketing. However, some positive developments can be observed in that respect. The possible development of FC Barcelona Park would strengthen the relation between the city and the club. Both actors may realise that they need each other to reach their objectives, in the same way as business companies are increasingly realising that they need to develop relations with their environment in order to safeguard their long-term competitiveness.

Notes

1   After the games, the venue functioned for a while as home to FC Barcelona's basketball team, but it suffered from overcapacity. Therefore, the team decided to return to its former home base (located on the Montjuïc as well).
2   An alternative for multi-unit facilities is the container concept: a building that consists of one large rectangular unit, which can accommodate temporary constructions for various activities.
3   In 1993 and 1994, the growth of tourism stagnated somewhat because of the worldwide recession.
4   All amounts in this chapter have been converted by the official Spanish peseta rate into Euros: 100 pesetas are equal to €0.60101.
5   Another incentive to keep the present structure is the fact that the tax rates for clubs and foundations are lower than the company tax rate.
6   One exception is the promotional video of Turisme de Barcelona, which contains some flashes of FC Barcelona.

Chapter Three

Helsinki

Introduction

Helsinki occupies an exceptional position in this study. The city's most successful football club – HJK – cannot be rated among such European top clubs as FC Barcelona, Manchester United, Feyenoord (Rotterdam) and Juventus (Turin). Nevertheless, the capital of Finland has been selected as object of a case study for several reasons.

Firstly, Helsinki was one of the candidate cities to host the twentieth Olympic Winter Games in 2006. Although in the end this event was assigned to Turin, it is still very interesting to analyse the objectives of Helsinki's bidding committee and the potential benefits for city marketing in particular. Secondly, the city has a lot of experience in hosting international (sporting) events, including the Olympic Summer Games in 1952 and various world and European championships. Thirdly, the city has been very active in stimulating and financing the construction of new venues, among which a multifunctional dome: the Hartwall Areena.

This chapter analyses Helsinki's efforts in the field of sports and city marketing. A short city profile is followed by a section that deals with Helsinki's Olympic ambitions. After that, investments in sports facilities are discussed and related to other developments in the sphere of sports. The subsequent section analyses Helsinki's city marketing strategy and the role of sports in urban management. Some final remarks conclude this chapter.

Profile

Helsinki is the capital and by far the largest city of Finland with a population of approximately 540,000. The Helsinki region is the leading economic area in the country in terms of trade, industry and commerce, education and culture. It boasts the largest concentration of people (more than 1.2 million) in Finland and includes the Helsinki Metropolitan Area, which comprises Helsinki, Espoo (190,000), Kauniainen (8,300) and Vantaa (166,000). Espoo presents itself as

Finland's High-Tech City. Of its industrial labour force, 73 per cent is employed by high-tech companies, including the Nokia Group. Helsinki's international airport is located in Vantaa. The region's share of the Finnish gross national product is 29.1 per cent (2006 Association, 1998).

The Helsinki region attracts many people from the rest of the country. Since 1950, the region's share of Finland's population has increased from 12.3 to 22.2, as depicted in Table 3.1. Moreover, Helsinki receives a relatively high number of foreigners. In 1997, 4.2 per cent of the citizens had a foreign nationality. In Finland as a whole, the share of immigrants is considerably smaller (1.4 per cent). However, the number of foreigners in Helsinki is still very small compared to other European cities.

Table 3.1   Population dynamics in Helsinki and its region

	1950	1970	1980	1996/1997
Helsinki	368,519	523,677	483,675	532,053
% of Finnish population	9.2	11.1	10.1	10.4
Helsinki region	496,517	827,400	930,368	1,137,244
% of Finnish population	12.3	17.6	19.5	22.3

Source: Helsinki 1997; Facts about Helsinki (Urban Facts, 1997).

The Helsinki region's labour market is to a large extent dominated by the Nokia Group. During the 1990s, this telecommunication company benefited from the rise of the information society and expanded its activities very rapidly. Today, Nokia's stock exchange value is unsurpassed in Europe and the firm employs about one-half of all graduates in the Helsinki region.

Hotel statistics show that tourism is a growth sector in the Helsinki region. In 1998, about 11 million tourists visited the region, including 2 million foreigners. The summer of 1999 broke all records, with 278,000 overnight stays a month, which equals 80 per cent of the available hotel beds (the average occupation rate is 68.2 per cent). Most hotels in Helsinki belong to the three- and four-star category, aimed at the business sector. The number of hotels for leisure tourists (family hotels) is relatively small. Many leisure tourists spend the night with families or friends. New hotels are under construction, but these are mainly focused on the lucrative business segment. Helsinki has a strong congress and exhibition infrastructure, including the Finlandia Hall (built in the late 1970s), the Fair Centre and several small congress centres.

Helsinki's Olympic Dream

The Finnish people are very sports-loving. There are over 6,500 sports clubs in Finland together counting 1.1 million members, including 350,000 children. From the number of active players, football is one of the most popular sports in Finland, especially among children and youngsters. Second in terms of participation is ice hockey, which is the number one sport in terms of spectators. Other popular sports are swimming, jogging, cycling, gymnastics, skiing, aerobics, tennis, skating and dancing. This affection for sports can be partly explained by historical facts. Until the Declaration of Independence (1917), Sweden and Russia had alternately ruled Finland. At the beginning of the twentieth century in particular, sport was a way to affirm the national identity of the newly independent Finnish state. There are close associations between Finland's Olympic success in the second and third decades of that century and the history of its national independence.[1]

In the 1930s, Helsinki started to run for the organisation of the Olympic Summer Games. After the city had lost once, the Games of 1940 were assigned to the Finnish capital. Many facilities – including the Olympic Stadium – had already been constructed when the outbreak of the Second World War threw a spanner in the works. The first post-war Olympic Summer Games (1948) were entrusted to London, but the subsequent Games (1952) took place in Helsinki.

The Olympic Games of 1952 contributed to Helsinki's economic development. First of all, several facilities for hosting (sports) events were constructed, such as the Olympic Stadium, a horse racetrack, a rowing track, a cycling track and an exhibition hall. Most of these venues are still in use today. Additionally, the event brought about improvements in the airport, the seaport, the sewage system and the rail infrastructure. Moreover, the Games gave the Finnish people a chance to show the world their capabilities for organising such an event.

Surprisingly, the Olympic Winter Games have never been organised in Finland, although this country seems to be an ideal location for winter sports. Until the 1990s, two Finnish cities (Lahti and Tampere) had tried to get the Winter Games, but without any success. In the beginning of the 1990s, the Finnish Olympic Committee took the initiative for a new attempt to bring the Winter Games to Finland. In contrast with the previous attempts, *Helsinki* was presented as the central location for the event, so that it would become the first city in the history to host both the Olympic Summer and the Winter Games. In the candidature file, the organising committee emphasised that Finland had been an active participant in the Olympic movement for more than 90 years and that Finland and Norway are the top two nations in the world in terms of Olympic medals won per head of population. Moreover, Helsinki had shown itself able to organise major events, such as the World Championships Athletics (1983 and 1994) and the World Championships Ice Hockey (1997).

The organising committee expected the Olympic Games to be of great importance for the further development of the sporting scene in Finland. The sports activities of children, young adults and mature sporting enthusiasts would be given a boost. From an economic point of view, the substantial regional and national growth benefits likely to accrue from the event in terms of employment have been estimated conservatively at €555 million[2] which is equal to 8,900 person-years (see Table 3.2). In view of the recession in the early 1990s, the main economic objective was to avoid losses. The benefits of 4,000 person-years generated in Helsinki would include approximately 2,800 person-years in direct employment growth. The indirect effects would amount to approximately 1,200 person-years.

In addition to these substantial contributions, the Olympic Winter Games would constitute a strong catalyst for enhancing the image of Finland and promoting Helsinki internationally (2006 Association, 1998). Because relatively few new venues were needed, a large share of the budget could be spent on marketing. Furthermore, the investments in traffic and other infrastructure which would be made for the Olympic Games were expected to engender improvements in productivity and environmental technology and greater accessibility of certain localities (2006 Association, 1998). Helsinki was aware that the construction of an Olympic Village would help to diminish the housing shortage and that the event could be a potential stimulus to building new small low-cost facilities for the local population.

Table 3.2   Direct and indirect impacts on employment derived from the 2006 Olympic Winter Games (person-years)

Helsinki	4,000
Helsinki region excluding Helsinki	800
Rest of Finland (including Lahti region)	4,100
Entire national economy	8,900

Source: Helsinki 2006 Candidate City; XX Olympic Winter Games Candidature File (2006 Association, 1998).

The budget for the bidding stage was €4.2m, from which three-fifths were financed by an unsecured loan from the City of Helsinki. In the case of success, the organising committee would pay this loan back. In the case of failure, the City of Helsinki and the Finnish state would each repay one-half of the loan. The remaining €1.7m would be collected from local partners. The modernised Olympic Stadium would be the focal point of the 2006 Olympic Winter Games, as it would be the stage of both the opening and the closing ceremony. Other sports disciplines that were to be staged in Helsinki are ice hockey, figure skating, speed skating, short-track speed skating and curling. All these disciplines can and would be organised in indoor venues. Four out of the five proposed competition sites were already present in 1999, when the International Olympic Committee had to choose the location of the event (see Table 3.3). The only venue that still had to be built, was the Myllypuro Speed Skating Arena.

Table 3.3   List of proposed competition sites for the 2006 Winter Olympics in Helsinki

Venue	Sport(s)	Present?
Hartwall Areena	Ice hockey	Yes
Helsinki Ice Arena	Figure skating and short track speed skating	Yes
Myllypuro Speed Skating Arena	Speed Skating	No
Pirkkola Ice Stadium	Curling	Yes
LänsiAuto Areena	Ice Hockey	Yes

Source: Helsinki 2006 Candidate City; XX Olympic Winter Games Candidature File (2006 Association, 1998).

Although it can be very cold in Helsinki, it is not an appropriate location for outdoor winter sports (events on snow instead of ice). Therefore, the organising committee had to find partner cities to stage the event. Lahti, located only 110 km north of Helsinki, was a logical first choice, because of its great experience in organising Nordic ski-discipline events, such as biathlon, cross-country skiing and ski-jumping. Consequently, the candidature file entrusts these events to Lahti, as depicted in Table 3.4. Nevertheless, the organising committee had to seek out a second partner city to host the remaining winter sports: the so-called alpine skiing events, including slalom and downhill. The organising committee had to cross the nation's borders to find that partner, since Finland lacks mountains high enough for professional alpine skiing events (the highest mountain has its peak at 1,324 m above sea level). Since 1990, the IOC has allowed certain events or disciplines of a sport to be organised in a neighbouring country. Because the Swedish Olympic Committee wanted to organise the Winter Games themselves,[3] the Norwegian Olympic Committee was contacted first.

It was not difficult at all to find a partner city in Norway, since the city of Lillehammer – which organised the Winter Games in 1994 – had all the facilities and the experience needed. Furthermore the organising committee decided not to build a new bob and luge track in Helsinki (or Lahti), because such a track was already present in Lillehammer (see Table 3.4).

Table 3.4　　Events in Lahti and Lillehammer

Lahti	**Lillehammer**
Biathlon	Alpine skiing
Cross country skiing	Snowboard
Ski jumping	Bobsleigh
Nordic combined	Luge
Freestyle skiing	

Source: *Helsinki 2006 Candidate City; XX Olympic Winter Games Candidature File* (2006 Association, 1998).

The municipalities of Helsinki and Lahti and the region of Lillehammer formed a bidding committee together with the national Olympic committees of Finland and Norway and the Finnish Ministry of Education (which is responsible for sports). Their joint bid was unique, because it would be the first Olympic Games to be staged in two countries.[4] In comparison with other bidding cities, the proposed budget of the Organising Committee of the Olympic Games (OCOG) was the lowest (US\$ 482 million), mainly thanks to the small capital investment. The non-OCOG budget was also very modest. This budget represents all investments related to the Olympic Games by local, regional or national institutes and the private sector that are not included in the official Olympic budget. In Helsinki's bid, this budget comprises the construction of sports venues and Olympic villages only. According to the bid-book there were no plans of public or private actors to invest in the airport, roads and railways or visitor accommodation. Obviously, public and private actors in Helsinki would have invested in facilities and infrastructure, but these investments were not part of the Olympic strategy as described in the candidature file beforehand. For instance, according to the candidature file, the number of hotel beds would increase from 6,255 to 7,166, but investment in hotel accommodation cannot be found in the OCOG or non-OCOG budgets.

The Fins and the Norwegians were optimistic about their chances, since EURO 2000 had been assigned to the Netherlands and Belgium and the World Football Championships 2002 to Japan and South Korea. For both events, the joint organisation is unique. They were also full of hope because they thought they could show the International Olympic Committee that all venues had already been constructed. That strength is emphasised in the official candidature file: 'By avoiding excessive new construction and siting activities in already existing competition sites and buildings, the 2006 Olympic Winter Games will conserve both energy and natural resources' (2006 Association, 1998). The people of Helsinki were mostly in favour of the Olympic Games. More than three-fifths supported the city's candidacy, while about 25 per cent were against the bid.

However, some unexpected developments have changed the rules of the game and possibly the chances of the participants. In 1998 and 1999, the International Olympic Committee became notorious, when members were accused of accepting gifts in return for a vote. Consequently, IOC members were forbidden to visit candidate cities and Helsinki could not invite them to visit the venues that were already present. Suddenly the organising committee had to change its strategy.

Helsinki's bid failed. In 1999, the IOC selected the Italian city of Turin to host the 2006 Winter Olympics (see Chapter Six). Some people argue that Helsinki has lost because the IOC wanted to avoid risks, such as a joint organisation by two countries. Unfortunately, IOC members do not have to explain their choice. Therefore, it is not possible to explain Helsinki's loss, or to find out what would have happened if the IOC crisis had not occurred.

Nonetheless, the bid has not been useless. It has helped to strengthen the sports network on a national level, because people and organisations (left and right wing groupings that had their own sports associations until 1993!) have been brought together. The bidding process, with its interactions among actors, has more or less streamlined the organisation of sporting events and bids in the future.

Investments in Sports Facilities

The city of Helsinki attaches much significance to the development of sports, which is considered essential to improve the quality of life. The municipal sports department is responsible for maintaining facilities and constructing new ones, as well as for marketing and providing information on sports services. The sports department is in charge of about 70 sports halls, including three skating rinks, five indoor swimming pools, two outdoor swimming pools, around 350 sports fields and one manège. In addition, the department supports nine indoor swimming pools operated by independent foundations. The department is also responsible for the maintenance of outdoor sports and leisure areas, such as tracks for walkers, joggers and cyclists (including 750 km of cycle paths and 180 km of skiing tracks). The Olympic Stadium is administered by a separate organisation in which the municipality is represented. Helsinki's budget for the maintenance and construction of sports venues is equal to the national budget (€16.8 million).

Moreover, the department subsidises amateur sports clubs. Citizens can choose from 600 sports clubs representing about 100 different sports. In 1999, the city granted activity and rent subsidies to nearly 400 sports clubs or similar organisations for handicapped persons or pensioners. One of the main objectives of the sports policy is to provide young people with social activities, in order to reduce crime. The city implements various projects designed for the unemployed, immigrants and socially excluded people. The sports facilities are equally scattered around the city to maximise their accessibility. Statistics show that this policy has been very effective. According to a recent study by the City of Helsinki, 67 per cent of the adults exercise three or more times a week and about 80 per cent of the children do some sort of exercise (not necessarily in a club). Almost 50 per cent of the youngsters between 13 and 18 in Helsinki are members of a sports club, which is very high compared to other cities in Europe and the rest of Finland.

Another objective of the sports policy is to use facilities for the socioeconomic revitalisation of deprived areas. The sports department is now building a new sports hall in one of the most deprived parts of the town (Myllypuro), which has high concentrations of unemployment (about 35 per cent), crime and social housing. The city believes that the sports hall will bind together several social groupings. At the same time, Helsinki aims to turn this area into a new urban growth pole by developing new business locations.

The Helsinki sports department finances not only amateur sports facilities but also professional ones. For instance, a new 50 m swimming pool that will accommodate the European Swimming Championships in 2000 is co-financed by the municipality (and by the national government). Two other venues that are co-financed by the City Council are the Hartwall Areena (a multifunctional venue) and the Finnair Stadium (football). Furthermore, the municipality has been responsible for legal aspects (approving construction plans) and infra-structure around these facilities. The municipality is usually not involved in the organisation of events (aside from the Olympics), as the national sports associations are mostly responsible for that. An example of an annual sports event is the Helsinki Cup, which has a great economic and promotional impact on the city. About 600 youth football teams from more than 30 countries participate in this event, which belongs to the five greatest youth events in the world. The city of Helsinki is also barely involved in stimulating the development of top-class athletes, which is the responsibility of the clubs and their associations. One exception to this rule concerns the special schools for top-class athletes, owned and managed by the city. Those schools are aimed at young people between the ages of 16 and 20. There are no similar institutes on the university level, however.

The Hartwall Areena[5]

The Hartwall Areena is located in the northern part of the city centre and is easily accessible by public transport thanks to the nearby train station. Its construction cost about €50 m, which has been partly financed by a €3m subsidy from the municipality. The remaining €47m came from a private investor: the Jokerit HC Group, a diversified, privately-owned entertainment and sports group. The core of this company consists of the ice hockey team Jokerit HC and Helsinki Hall, the company operating Hartwall Areena. In 1999, a football team, FC Jokerit, was added to the group.

To be on time for the Ice Hockey World Cup, the municipality accelerated the winding up of procedures. The multifunctional venue opened its doors in 1997. It can be used for various activities, since its flexible concept makes its possible to transform a concert hall into a complete basketball court within 24 hours, or to asphalt an ice rink and turn it into a full-scale carting track within a few days.

The Areena is provided with an advanced information and communication infrastructure, including optical wires. The Hartwall Areena has a seating capacity of up to 14,000, including 4,000 restaurant seats and 78 skyboxes (of which two have a sauna with view of the hall!). The venue is the home ice rink of Jokerit HC, one of Finland's most successful ice hockey teams. On average, 12,000 people attend the home games (30 per year). The Hartwall Areena complex also contains a practice hall, built 20 m into a rock next to the main arena. It serves as practice rink for Jokerit HC when the main arena is reserved for other events.

During the first 12 months after its opening in April 1997, the Areena welcomed 1.5 million spectators, attending 220 events (Jokerit HC Group,

1999). Table 3.5 contains a list of some major events that have been accommodated by the Hartwall Areena. Today, the Areena yields a profit, which is quite unique for a sports venue.

Table 3.5   Some major sporting events in the Hartwall Areena

Helsinki International Horse Show
Volvo World Cup Final 1998
Strongmen Helsinki Grand Prix
European League basketball games
Thunder Indoor Carting
World Figure Skating Championships 1999
Professional World Aerobic Championships 1999

Source: Hartwall Areena (Jokerit HC Group, 1999).

The Finnair Football Stadium

In August 2000, Finland's first 'real' football stadium will be opened in Helsinki. The 11,000-seater venue is located next to the Olympic Stadium and possesses all conceivable facilities such as skyboxes and business seats. If football becomes more popular in Finland, the capacity can be extended to 25,000 seats by building new stands. The stadium will be equipped with a high-tech communication and information system similar to that of the Hartwall Areena. It will be the home of two football clubs: HJK and FC Jokerit. Currently, these teams draw on average 2,000 spectators per home match. The national team, which draws on average 24,000 spectators per match, will play most of its games in the Olympic Stadium (with a capacity of 36,000). The €18.5 million investment is partly financed by the private sector (€11.8 million) and partly by the public sector (€5.9 million). Jokerit HC Group will rent the venue for 10 years and has paid the rent in advance (€0.8 million). The new football stadium will be owned by the city of Helsinki.

HJK Helsinki is the most successful club in Finland, being the only one ever to have played in the lucrative Champions League (the winner of the national league has to play three qualification matches to enter this tournament). In that memorable season (1998–99), HJK earned twice its yearly budget. European top clubs bought many players from the team. Although football is the most popular sport in Finland in terms of the number of active players, in terms of advertising and television viewers, professional football is not as important as ice hockey, basketball, volleyball, skiing and swimming. Nevertheless, football enjoys growing attention from the media and business (for sponsorship and advertising), stimulated by the increasing popularity of European competitions (with or without Finnish clubs).

The Finnish Football Association attaches great importance to the social value of football. Accordingly, a special programme has been launched, called 'all stars'. The programme is based on the idea that everybody should be able to participate in football: football as a way to spend leisure time. If necessary, rules can be relaxed to some extent. Moreover, the concept comprehends smaller teams (with five or seven players in each team). The city of Helsinki is not involved in this project.

Helsinki's (City-marketing) Strategy

From the preceding sections can be concluded that Helsinki has a relatively strong sports infrastructure, including the facilities that were constructed for the 1952 Olympic Games, the Hartwall Areena, several facilities that would have been used for the 2006 Winter Olympics (see Table 3.3) and the Finnair stadium (to be opened in August 2000). Furthermore, Helsinki has hosted several World Championship, European Championship and World Cup events. Potentially, sports can play a great part in Helsinki's city-marketing strategy. To analyse the relation between sports and city marketing in Helsinki, we need to discuss Helsinki's (city-marketing) strategy first.

The Helsinki City Council has not (yet) defined a city-marketing strategy. However, two strategic policy documents contain city-marketing elements. The first document is the internationalisation strategy – written in 1994 – aimed at developing Helsinki into a strong centre serving the Nordic countries, the Baltic states and northwest Russia. To acquire that status, the following priorities have been set:

1 to develop Helsinki as a centre of education, science and research;
2 to improve the opportunities for the business sector in the city;
3 to develop better external transport and communication links;
4 to enhance the city's cultural profile;
5 to improve the quality and pleasantness of the urban environment in the city;
6 to develop Helsinki as a city providing a wide and versatile range of welfare services;
7 to boost the city's international marketing and develop international cooperation networks; and

8   to prepare the city for any actions necessitated by possible Finnish
    membership of the EU.

After 1994, several projects related to these areas were carried out. The
candidacy for the Olympic Games and the construction of the Hartwall Areena,
are both examples of sports-related investments that fit in with the objectives
to enhance the city's culture profile (4) and to boost the city's international
marketing and develop international cooperation networks (7).

The second strategic document is an update of the internationalisation
strategy. In 1999, the City Council evaluated the first strategy and defined
new policy priorities for Helsinki's international activities, derived from the
city's strengths and opportunities as well as the target set for its international
position (Helsinki City Council, 1999). In comparison with the first strategy,
more policy priorities are related to promotion (instead of product
development):

1   to *promote* foreign investment in Helsinki, to assist the internationalisation
    of local business and develop Helsinki as an attractive city for international
    companies;
2   *to strengthen Helsinki's position* as a venue for urban tourism and a host to
    conferences and congresses;
3   to *promote* the hosting of internationally important cultural events and
    activities;
4   to *promote* the organisation of internationally attractive sporting events
    suited to Helsinki;
5   to ensure the preservation of the city's natural values and green environment
    and *to emphasise these in marketing the city*; and
6   *to promote and market* Helsinki as a centre of academic expertise and
    research and strengthen cooperation between the city and its institutes of
    higher education.

It could be concluded from the policy shift that the City Council had
come to realise that an attractive city needs to be promoted in order to attract.
This can be illustrated by the explanation of the objectives to promote the
hosting of internationally important cultural events and activities and
internationally attractive sporting events suited to Helsinki: the considerable
cultural (and sports) infrastructure which Helsinki has created for itself (product
development) provides the essential framework for cultural creativeness.
Following the construction of this infrastructure, the main attention will now

be devoted to putting it to use, i.e. the content and functioning of the buildings (Helsinki City Council, 1999).

This can be interpreted as a growing need for city marketing, which is more or less confirmed in the strategy itself:

> Helsinki's distinctive cultural, historical and geographical features and the above-mentioned competitive factors all influence the city's chances of successfully competing for investments, jobs, tourists and congresses. Marketing is a means to bring out and make known the city's special features and competitive factors (Helsinki City Council, 1999, p. 17).

Presenting the city at international (sports) events is one effective method to market the city and project a positive and recognisable picture of the city (brand marketing). In other words: it is necessary to increase the awareness of Helsinki's qualities to enhance its image. At present, the city has no image at all among some people (they simply do not know the city) and the image of a cold and expensive city located near the polar circle among others. Few people are familiar with Helsinki's qualities such as its extensive cultural and sports infrastructure, its good accessibility (see Table 3.6), its relatively mild temperatures (the hottest month is July with an average maximum temperature of +19°C) and its beaches.

Table 3.6   Flight times (in hours) to Helsinki

St Petersburg	0.45
Moscow	1.30
Berlin	1.30
Brussels	2.30
London	2.30
Paris	2.30
New York	7.00
Tokyo	10.00

Source: Helsinki Region: A European province with prospects and potential (Uusimaa Regional Council, n.d.).

Events

It is not clear (yet) what strategy the city will use to promote the organisation of internationally attractive sporting events suited to Helsinki. As described above, the City Council is not usually (financially) involved in the organisation of events. On the one hand, possible deficits have no consequences for the urban budget. On the other, the city has little influence on the kind of attractions that are created, which makes it difficult to attract events to Helsinki that fit within the city-marketing or event strategy and contribute to the socioeconomic development of the city.

Another complicating factor is Helsinki's exceptional position within Finland as by far the largest city of the country. Consequently, it has a monopoly in the national market for big (sports) events. Except for skiing events (usually organised in Lahti), all international sports events attracted by Finnish sports associations take place in the capital city. The national government always supports Helsinki's candidacy, unlike the national governments of other European countries – such as Great Britain, Spain, Italy and the Netherlands – where several cities have enough critical mass to run for major sporting events. Consequently, Helsinki's main concern is not to attract events, but to be select events *suited to Helsinki*. The need for selectiveness is growing because the city is slowly becoming congested in the popular summer months. In June, when school holidays start, hotels are fully booked up and conference organisers complain that they have to accommodate their guests in villages more than 100 km from Helsinki. Furthermore, the increasing competition between events reduces the average number of participants and spectators. That tendency is particularly worrying, since the near absence of congestion used to be a major strength of Helsinki.

City Promotion

The second internationalisation strategy emphasises the importance of promotion. Currently, two organisations are responsible for city promotion. The first is the Tourist Office, which is mainly engaged in providing information. It has a staff of 11, including desk employees, and a relatively small budget (€1.2 million).[6] The Tourist Office informs visitors of cultural and sporting events to take place in the near future. It appears difficult for the Tourist Office to use sports events in city promotion, because event organisers usually want to do the marketing themselves. Moreover, public bodies have to deal with sponsoring companies of (sports and cultural) events that prefer branch-exclusiveness. That situation makes it very hard for the city to publish promotional brochures that give a review of all events. One option to solve this problem is to oblige event organisers to give basic event information if they want to receive the permits they need from the city to organise the event.

The second city promotion organisation is the Information Bureau, which has a network of offices in 10 foreign countries. The main task of these offices is to provide information and undertake marketing activities such as press journeys. The Information Bureau has also been involved in the Olympic bidding process. It cooperates with other municipalities in the region.

Conclusions

Primarily, Helsinki's bid for the Olympic Winter Games was directed at getting a sports event to Finland. Indeed, the sports event as such was an objective in its own right. That is evident from the fact that the bidding committee claimed the event on the basis of (among other things) the sportsmanship of the Finnish people. That the Finns are very sports-minded is illustrated by the percentage of active sportsmen and women and the international successes of Finnish athletes.

However, the event was regarded not only as an objective but also as an instrument with which to realise other objectives. The city of Helsinki was very much aware of the potential benefits of the Games and the opportunities to promote the town as a sports city in particular (brand marketing). Nevertheless, there was not (yet) a clear vision on how to optimise those benefits. The bid included some investments in the sports product of Helsinki, but specific product development was not one of its main objectives. In fact, the organising committee set great store by the sustainability of the event, minimising the construction of expensive venues and avoiding overcapacity in the supply of sports products. That defensive strategy – which also reveals itself in the low non-OCOG budget for investments – may be ascribed to the poor economic situation in the early 1990s. One advantage of this strategy is that it minimises the chance of financial problems, such as budget deficits of the organising committee and the local government. Obviously, the fact should be taken into account that this conclusion is based only on the candidacy file. If Helsinki had won the bid, the city would possibly have changed its strategy. Despite the loss, the city seems to have profited from the bidding process, since it has contributed to the creation of strategic networks on the national and local level.

In the latest update of the so-called international strategy, the city clearly shows its ambitions to use sports for city marketing in the broadest sense. In the course of time, the city has obtained a competitive advantage in the field of sports by constructing venues and gaining experience in organising big events. Moreover, the city is relatively clean, safe and internally accessible, compared to other cities. One of its few weaknesses is the absence of a European top-level football club. The city has decided to subsidise the construction of a football stadium, but that is not a guarantee for success in sports. Nevertheless, Helsinki has most of the ingredients to promote itself as a city of sports, especially because of its excellent facilities for 'on the street' sports (running, jogging, cycling). Tne city has shifted the emphasis from product development to product promotion and general city promotion (branding).

Helsinki wants to attract sports events to use them for general city promotion, which can help to give the city a better image (especially in terms of external accessibility, price level and climate). However, to optimise the synergies between sports and city marketing, the city and other relevant actors face a number of challenges. First, the city should define a city-marketing strategy, preferably in cooperation with surrounding municipalities, local businesses and the local population. Second, there is a need to define a common vision on the role of sports in marketing the city, based on (among other things) the knowledge of event organisers, clubs and venue operators. Helsinki is confronted with the challenge of strengthening the mutual dependence between producers of the sports product and the city. There is a growing need to define an event strategy, in order to avoid spatial problems for the city (congestion) and financial problems for the event organisers (who suffer from too much competition).

Furthermore, the optimum use of sports for city marketing requires a strategic city-marketing organisation: a partner and facilitator for actors in the sports cluster. The Tourist Office and the Information Bureau are mainly engaged in operational city marketing activities, both being focused on promoting the city and providing information to visitors (Tourist Office) and press and business companies (Information Bureau). The sports department is not involved in marketing the city. Therefore, the city needs an organisation that functions as spider in the web to guide investments in the sports product (the construction of venues, subsidies for clubs and the acquisition of events) and stimulate the use of sports as an instrument for city promotion (external target groups) and social revitalisation (internal target groups) leading to economic growth (employment).

Notes

1   According to Mr Tapani Ilkka (Chairman of the Finnish Olympic Committee) in one of the introductions to the 2006 Winter Olympics candidate file (2006 Association, 1998).

2   All amounts in this chapter have been converted by the official Finnish Mark (FM) rate into Euros: 100 FM is equal to €16.818.

3   In the end, the Swedish Olympic Committee withdrew from the competition.
4   In fact it would be the second time, since the equestrian sports of the Olympic Summer Games of Melbourne (1956) were organised in Stockholm.
5   Source: www.hartwall-areena.com.
6   In 2000, the budget will be raised to €1.5 million.

Chapter Four

Manchester

Introduction

For several reasons, Manchester is an interesting case in the context of sports and city marketing. In the first place, this industrial city is known all over the world as the home base of one of the most famous football teams: Manchester United. Although the club's stadium is actually located outside the city borders (in the borough of Trafford), United is largely responsible for the sports image of the city. The second football club, Manchester City, is also worth analysing, because it is located within the city borders and has a very good relationship with the City Council. Secondly, the city has gained a lot of experience in bidding for and organising big events, such as the Olympic Games (two bids without success), the Commonwealth Games (to be staged in 2002) and the European Football Championships (1996). The bidding processes have proved to be very valuable for the city. Thirdly, Manchester actively uses sports as a lever for city marketing, not only in terms of promotion but also in terms of social and economic revitalisation. Indeed, Marketing Manchester, the body responsible for marketing the city, explicitly aims to use sports as a tool for city marketing (Phelan, 1997). Furthermore, the city pays much attention to the legacy of events and the participation of youngsters in sports. Finally, the metropolitan region of Manchester shows a clear concentration of sports-related industries, which can also be interesting from a marketing point of view.

This chapter is organised as follows. After sketching the city and its history, we will discuss the legacy of bidding for sports events. The next section analyses the city's efforts to maximise the legacy of the Commonwealth Games, followed by a debate on the relation between the city and the sports clubs (including Manchester United). The final section concludes.

Profile

The city of Manchester is the capital and commercial, educational, medical and cultural centre of England's northwest region, the second largest economic region in the United Kingdom (UK) outside London. The city has a population of about 400,000 people, and the metropolitan region (Greater Manchester) accounts for more than 2.5 million. The region comprises the cities of Manchester and Salford and the Metropolitan Boroughs of Bolton, Bury, Oldham, Rochdale, Stockport, Tameside, Trafford and Wigan.

Manchester is internationally known as the first city of the Industrial Revolution. Like many other industrial cities, Manchester has suffered from the decline of the industrial sector, resulting in severe social problems. Between 1972 and 1984, unemployment in the industrial sector declined by about 207,000 jobs (Phelan, 1997). Nowadays, about 70 per cent of the working population is employed in the service sector, which has experienced an increase over the last two decades. However, the area retains a strong industrial and manufacturing base, employing more than a fifth of the working population (300,000 in the Manchester area and almost 700,000 in the wider region) (see Table 4.1). About 60,000 people are unemployed, which is 4.8 per cent of the economically active population.

Table 4.1   Employment in the Manchester TEC area, 1995

Sector	Employment	%
Agriculture	230	0.0
Energy and water	4,650	0.9
Manufacturing	81,030	16.0
Construction	18,860	3.7
Distribution, hotels and restaurants	102,170	20.2
Transport and communications	41,530	8.2
Banking, finance and insurance	104,140	20.5
Public administration, education and health	136,480	26.9
Other services	17,860	3.5
Total	506,950	100.0

Source: *AES* (Office for National Statistics, 1995).

Manchester is part of a group of cities (together with Birmingham, Edinburgh and Glasgow) that are considered 'second cities' in the UK (London being the first city). They share the challenge of finding their own competitive niche in the British urban system. Manchester uses sports and culture as distinguishing elements (its region has the largest theatre sector in the UK and a wealth of art galleries, museums and libraries), but the city is also an important financial and educational centre. There are more than 15,000 people employed in banking and finance and more than 60 banking institutions. Students can choose from four universities in the city and one in neighbouring

Salford. Manchester Metropolitan University is one of the UK's largest universities and has established extensive links with industry and employers. One of the centres of expertise at this university is related to the topic of Sport Science.

Thanks to its international airport (which offers 175 destinations), the city is very accessible. A new two-mile second runway was completed in the year 2000, and by 2005 the airport expects to have doubled its capacity, handling over 30 million passengers a year. Furthermore, the city has three national rail stations. Recently, the internal accessibility has been improved by MetroLink, a new tram system.

Tourism has always been a strong industry for Manchester and its region (the northwest) and is likely to grow more so as Manchester's international profile has soared. In 1995, Greater Manchester received 2.8 million visitors from the UK and 520,000 overseas visitors. Tourism employed 61,000 in 1991 (Census of Employment). The city also attracts many business tourists. Manchester has become the largest convention market outside London and a significant international location. In 1995, more than 17,000 conferences and meetings were held at 31 venues.

The Legacy of Event Bidding

Today, Manchester is very actively bidding for and organising sports events. Until the mid-1980s, however, Manchester had no reputation as a sports city. It lacked experience in organising sports events and its sports infrastructure was relatively weak. What has happened since then?

In the late 1980s, the National Olympic Committee of the United Kingdom wanted to nominate one city to participate in the worldwide competition for the 1996 Summer Olympic Games. Soon, the media were convinced that London (host of the Games in 1908 and 1948) would win that race. The chances of Birmingham (the second candidate) were considered poor, since that city had lost the bid for the 1992 Olympics. Suddenly, a third candidate city put itself forward: Manchester, a remarkable candidate in several respects. First, the city had no experience in organising sports events and at the time lacked an adequate sports infrastructure. Moreover, the bid was not initiated by the city, but by a private entrepreneur: Bob Scott, a theatre director. The bid was entirely promoted by the private sector, the involvement of the public sector being limited to giving the green light. Another remarkable aspect was that the Olympic Stadium was planned outside the city borders. In spite of all these disadvantages, Manchester succeeded in beating London and Birmingham. As a consequence, Manchester received a lot of media attention during the international competition for the Olympic Games, which was won by Atlanta in the end. Nevertheless, the bidding process had raised the city's profile amongst the international and sporting community (Phelan, 1997), and increased the number and frequency of interactions in local networks.

After the loss, Manchester did not lose heart. On the contrary! The city decided to make a new bid for the 2000 Olympic Games, this time with active backing from the City Council. It was a private-sector initiative but with public support secured. Again, Manchester was selected, leaving London, Birmingham and Sheffield behind. This bid was better resourced, more strongly supported and more actively promoted than the first. A strong public-private partnership was developed, which created cooperation among areas and established agencies, and which has remained a major force in the resurgence of the city. This time, the city considered the Games a potential catalyst for urban revitalisation, being inspired by Barcelona's strategy for the 1992 Olympic Games. The bid was based on the marriage of urban regeneration to the provision of sporting facilities. This time, facilities were planned *inside* the city borders instead of outside. The Olympic Stadium would be constructed in East Manchester, an industrial town quarter with many desolate or outdated factories and high unemployment figures (30 to 50 per cent). The national government was very supportive and held out the prospect of a £17 million subsidy. The stadium was planned to be the heart of Sportcity, including many other venues.

Two sports facilities were built during the second Olympic bidding procedure: the National Cycling Centre (the first indoor cycling venue in Great Britain) and the Manchester Evening News Arena (to stage basketball). The National Cycling Centre – the Manchester Velodrome – was developed as a joint venture of the English Sports Council, Manchester City Council and the British Cycling Federation. It is located in the Sportcity area (see next section). The central event area has permanent seating for 3,500 spectators, hospitality boxes and facilities for VIPs, officials and media. It also provides management accommodation, competitor services, spectator services, technical services and car and coach parking. It hosted the 1995 Cycling World Cup and the 1996 World Championships. It is recognised as one of the world's fastest tracks.

The 20,000 capacity Manchester Evening News Arena was built alongside Victoria Station, affording direct access to both the MetroLink trams and regional railways; with a bus depot and taxi rank adjacent to the station, all forms of public transport are easily accessible from the arena. The construction of the Arena was already planned before the Olympic bid. Costing £56 million to construct, the Arena – the largest indoor entertainment centre of its type in Europe – is co-financed by a £35.5 million subsidy from the central government (the largest-ever city grant) and a £2.5 million donation from the European Economic Development Fund. It is a multifunctional venue, accommodating sports events, music concerts and conferences.

The Arena is home to the Manchester Giants basketball team and the Manchester Storm ice-hockey team. With an average of 7,000 spectators a match, Manchester Storm is one of the most popular ice-hockey teams in Great Britain. The team holds the European ice-hockey attendance record (more than 17,000 spectators!). The organisers of the (British) Icehockey Superleague have decided to move the annual Ice Hockey Finals from Wembley Arena to the Manchester Arena. In view of the city-centre location of the Arena and the number of people who have attended them, the ice-hockey matches have undoubtedly generated income for local businesses. In recognition of this, the club was awarded the Greater Manchester Business Through Tourism Award. Another positive element of the club's impact on the region is the considerable number of youngsters who have been inspired to establish their own roller-hockey teams.[1] The Manchester Giants play on British top level and draw approximately 4,000 spectators a match, with a peak of more than 14,000, which is a UK attendance record.

The city perceived the potential benefits of the Games to the economic and social structure of the city, the promotion of the region and its image and to community pride. It was recognised at an early stage that these benefits could only be maximised through coordinated positive action. Manchester harnessed the city's most representative social and economic institutions for the express aim of continuing the economic momentum that the dynamics of the bidding process had set in motion (Phelan, 1997). Some benefits that were identified are: improving and investing in the region's sporting facilities; creating new jobs and sustaining existing employment; improving the region's housing and infrastructure; creating new development opportunities; opening up new business opportunities; generating income in the local area; enhancing the environment; promoting the region and its image; attracting inward investment; creating new training opportunities; delivering social and cultural benefits to the community; enhancing higher education and community pride. Obviously, Manchester aims to use the event for city marketing in the broadest sense of the word.

Although Manchester did not win this bid either (the Games were assigned to Sydney), it was a 'race worth losing'. First, Manchester's international profile has been boosted immeasurably. In the week before the voting, the city received 10 times more attention from the world's press than any other city in the United Kingdom. Secondly, the bidding process increased cooperativeness among the various actors in the city and improved the relationship between the city and the national government. Thirdly, it resulted in improved strategies for transport, telecommunications, security, tourism and more. Fourthly, the bid brought civic pride and new confidence to the city. Fifthly, it generated financial benefits to the sum of £200 million, including capital investments (£75 million) from the central government. The benefits of bidding are confirmed in a report of the National (British) Heritage Committee, which states that:

> bids to stage major sporting events ... can operate as a catalyst to stimulate economic regeneration, even if they do not ultimately prove successful ... Once the initial redevelopment has taken place [referring to Sheffield and Manchester] the existence of high-quality facilities means that the cities concerned are able to attract other sports events. The impact does not stop here. Many of the facilities are suitable for other uses such as conferences and concerts. In addition, the favourable publicity which can follow from a successful event may increase the attractiveness of a city, raise its profile overseas, and enable it to attract an increasing number of tourists (House of Commons: National Heritage Committee, 1995/Sports England, 1999, p. 26).

Objectives of the Commonwealth Games

The Commonwealth Games

After the failure of the second Olympic bid, Manchester decided to run for another event, the Commonwealth Games, in order to realise its revitalisation plans, including the construction of Sportcity. The Commonwealth comprises Great Britain (England, Wales, Scotland, Northern Ireland) and former British colonies, such as Canada, Australia, India, Nigeria, South Africa and Pakistan. The Commonwealth Games take place every four years in one of the 70 countries of the Commonwealth. It is a multi-sport event, including (among other sports) athletics, gymnastics, swimming, cycling and boxing. In 1995, the Commonwealth Games 2002 were entrusted to Manchester. According to Phelan (1997), the bidding process has already brought significant economic and social change, and the international image of Manchester and the northwest has been enhanced considerably. With an expected television audience in excess of 500 million, it is the biggest multi-sport event to be staged in the UK since the 1948 Olympic Games.

The focal point of the event will be Sportcity, an area of 40 hectares located in the eastern part of Manchester. At the heart of the complex is the new City of Manchester Stadium, the new home base of Football Club Manchester City. The Opening Ceremony for the Games will be held in this venue with a 38,000 capacity. The stadium will not be *the* National Stadium, a title that has been awarded to London's Wembley. Alongside the stadium, there will be a new squash centre (the National Squash Centre) and an athletics arena with an additional outdoor warm-up track. The existing National Cycling Centre is also a part of Sportcity. Five of the 17 Commonwealth sports will take place within Sportcity: athletics (Stadium), cycling (National Cycling Centre), squash (National Squash Centre), table tennis (Indoor Table Tennis Initiative) and rugby (Stadium). Moreover, the area will be the heart of the Games, hosting nightly festivals and entertainment for the one million people expected. Other venues that will be used are the Manchester Swimming Pool Complex (aquatics), the Bolton Arena (badminton), Heaton Park (bowls), Wythenshawe Forum (boxing preliminaries) the Manchester Evening News Arena (boxing finals), the G-MEX Centre (gymnastics, judo and wrestling), the National Shooting Centre (shooting), the Royal Northern College of Music (weightlifting), Salford Quays (triathlon) and Carrington (hockey).

The Manchester Swimming Pool Complex (which was opened in September 2000) is located south of Sportcity, in the university campus area. This facility is to be developed by the Manchester City Council and the three Manchester universities. It will provide Manchester with a much-needed 50 metre pool right in the city centre. The English Sports Council has granted £21.9 million (from the National Lottery resources) towards the £32.2 million construction costs, the balance being funded by the City Council and the three universities. The proposed complex will contain one 50 metre, eight-lane pool for swimming and water polo competitions, one 50 metre, four-lane training pool; a 25 metre pool for diving and synchronised swimming, a leisure pool with flumes, water slides and bubble pools, and health and fitness areas. Permanent seating for 1,500 people will be temporarily increased to around 2,500 for the Commonwealth Games. The complex will meet the needs of competitive as well as recreational swimmers. According to the City Council, the venue will enhance the sporting infrastructure and strengthen Manchester's status as a sporting city (Manchester 2002 Ltd, 1998).

City Pride

The Commonwealth Games are part of an integral vision on metropolitan development. This vision is outlined in the 1994 City Pride Prospectus, produced by the Manchester City Pride Partnership (in which the cities of Manchester, Salford, Trafford and Tameside participate). In that vision, Manchester has to be developed into a European regional capital and a centre for investment growth; an international city of outstanding commercial, cultural and creative potential and an area distinguished by the quality of life and sense of well-being enjoyed by its residents (City Pride Partnership, 1994). In view of the change of government in 1997 and the New Deals offered across education, health, employment, training and regeneration, the Partnership decided to add a fourth objective in the second City Pride Prospectus: to develop the region into an area where all residents have the opportunity to participate in, and benefit from, the investment and development of their city and therefore live in truly sustainable communities (City Pride Partnership, 1997).

The Partnership has identified four key policy areas through which they aim to achieve these objectives and realise the vision. One of these policy areas refers to the Commonwealth Games. The partners recognise that this event provides unique opportunities to give impetus to a range of initiatives; investments in people, sports participation and excellence, economic growth, cultural development, and promotion of Manchester and its region. The main objective is to secure maximum *long-term* benefits. To reach that objective, integrated cross-agency and multi-agency initiatives are needed, since problems and opportunities do not confine themselves to administrative boundaries or organisational parameters.

In the 1994 Prospectus, sports and culture were already considered important employers and wealth generators in the Manchester area. Sports and culture can be used to enhance the image, reputation and marketability of the City Pride area. Manchester belongs to a group of British cities (together with Birmingham, Sheffield and Glasgow) that view sports as a powerful tool to enhance the physical fabric of communities, to stimulate the local economy, and to improve its image with outside investors and tourists (Sport England, 1999). Moreover, the Commonwealth Games and the years leading to 2002 provide an opportunity for the creative sector (music, theatre, art) to showcase its strengths. The City Pride Partnership has developed a number of projects to capture the potential long-term benefits arising from the event. The Commonwealth Games Organising Council has agreed to establish an Executive Team to harness such efforts throughout all the subregional groupings in the entire northwest region (including the City Pride area). To maximise benefits and ensure cross-fertilisation, traditionally discrete programme areas (e.g. economic, sports, culture) will be integrated through a comprehensive Legacy Programme with five core themes (City Pride Partnership, 1997):

* *economic growth and development*: the Manchester Investment and Development Agency Service (MIDAS) is overseeing the development and implementation of a strategic framework and a detailed action plan, which builds on the expertise of all the relevant local and regional organisations, and involves the private sector in taking initiatives forward; the Games will create employment (in the construction and exploitation stages), contribute to regeneration in East Manchester and the Piccadilly Gateway and will possibly result in more trade and investments. One project that has been identified concentrates on developing trade (principally exports from the UK) with all Commonwealth Countries. The MIDAS programme includes such activities as developing new networks and maximising the impact of existing ones with Commonwealth countries, holding seminars and business events, trade missions and events in target countries (Malaysia, Singapore, Australia, Canada, South Africa and the Indian subcontinent), placing a northwest representative in these countries and the development of focus groups in key market sectors (creative industries, health industries, sports-related industries, information and communication technology, environmental industries, aerospace, automotive production and tourism);[2]
* *social regeneration*: investing in people, social cohesion and community capacity; through the Social Volunteer Programme, the City Pride partners are determined that local people will have access to the core Commonwealth Games Volunteers Programme; at the same time the event offers a unique platform to raise health awareness in the region, through the realisation of the sporting and fitness opportunities available; the strategic projects make use of existing participative channels such as schools, youth clubs, community centres and training classes. One of the strategic projects that have been identified is the Street to Stadium Project. This project is directed by a Trust with representatives from the three original City Pride local authorities, the Greater Manchester Police, the Probation Service, the universities, the Sports Council and the Youth Charter for Sports, in addition to many community sports groups and the private sector. The Trust recognises the significant role to be played by sport, and aims to provide young people living in the inner City Pride area with opportunities they

may not otherwise be able to afford, such as access to training, education and facilities, in order to develop previously untapped potential and hidden talents in sports, the arts and cultural activities;

- *sports*: developing community participation through sporting excellence at an international level; the Games will provide the impetus to raise the standard of sports at all stages, creating new sports clubs and sporting opportunities, giving local children and young people a greater belief in their ability to participate in sport at the highest competitive levels. Thanks to the construction of new facilities, Manchester will be able to run for more events, after the Games;
- *artistic excellence and cultural diversity of international repute*: the event will provide an opportunity to showcase Manchester, conveying to a worldwide audience the qualities of the city and its culture;
- *promoting Manchester and the northwest*: the event is an instrument to create and promote a special image within a structured marketing strategy. As such, the Games are part of a longer-term strategy to promote the city and wider region as an area of sporting excellence, with regular national and international events. A coordinated events and bidding strategy (concerning sports and cultural events) will be developed, and will build upon and involve local capacity for promoting events through partnership. One of the strategic projects will deliver an integrated visitor service for tickets, accommodation bookings and transport, and for the dissemination of visitor information.

Some of these objectives can also be found in the bid for the Single Regeneration Budget (SRB) programme, which is focused on the northwest region. It aims to tackle socioeconomic issues such as low education levels, (high concentrations of) unemployment, social exclusion and disaffection, and the need to develop target growth sectors of the northwest economy. The main vision of the 2002 North West Partnership is to create social and economic benefits for the Manchester region by harnessing the opportunities and maximising the benefits arising from hosting the Games. In particular the Manchester region aims to ensure that disadvantaged and Commonwealth-originating communities have fair access and equity in sharing the benefits arising from the Games. Three strategic objectives have been identified which will deliver the vision:

- *skills and education*: to improve the skills; educational attainment and personal development within targeted disadvantaged areas,

Commonwealth-originating communities and young people, by harnessing the opportunities, interest and fascination created by the Commonwealth Games;

- *community and health development*: to create greater cohesion and improve skills within targeted disadvantaged and Commonwealth-community groups through participation in celebratory events and health improvement programmes linked to the games;
- *business competitiveness*: to use the commercial opportunities generated by the Games to improve the competitiveness of small and medium-sized enterprises (SMEs) in targeted sectors and ethnic minority businesses in the region.

An overview of the strategic objectives, programmes and target groups is presented in Table 4.2.

According to the bid, the 2002 Commonwealth Games presents several opportunities to be linked with regeneration strategies on national, regional and local levels (The 2002 North West Partnership, 1999):

- *image*: transmit a new and vibrant image – this will link with and support trade, inward investment and tourism strategies;
- *competitiveness*: create new investment, trading and commercial opportunities; this supports competitiveness policies at regional and national levels;
- *education and employment*: develop regionwide approaches and programmes to allow the region to act in concert to enhance educational attainment and skill levels, to improve the labour market and reduce unemployment; this links with both national and local policies on education and employment;
- *social inclusion*: accelerate programmes for promotion of healthy active lifestyles including participation in sport, and support self care to reduce ill-health. This will link with national policies in health and sport;
- *cultural creativity*: use the lead-up to the Games, the event itself and its spin-off activities, to exploit the worldwide exposure of the region and showcase its cultural diversity, creativity and strengths; this links with recent policy development on creative industries;
- *community cohesion and diversity*: demonstrate the vibrancy, importance and contribution of Commonwealth-originating communities to the northwest – this area of work is important to national and local policies, both on culture and on media, and on race equality;

Table 4.2 Strategic objectives, programmes and targets of the SRB programme

Strategic objectives	Skills and education	Community and health development	Business competitiveness
Programmes	Volunteer programme Passport 2002 Regional Grid for Learning	Let's Celebrate Healthier Commonwealth Communities	Prosperity North West Programme
Targets	Individuals: school children young people unemployed underemployed	Community groups: disadvantaged areas Commonwealth-originating communities voluntary sector groups	Commonwealth countries: NW target sectors (tourism, creative industries, health, sport, environmental, ICT, aerospace, automotive); ethnic minority SMEs across the northwest

Source: The 2002 NW Economic and Social Programme 1999–2004 (The 2002 North West Partnership, 1999).

- *civic pride and celebration*: generate pride and community spirit in sharing the passion and success in hosting such a unique, once-in-a-lifetime event.

In 1995, the Manchester City Council developed a sports strategy to maximise the potential of the Commonwealth Games. This strategy has three pillars: facilities, events and participation. The City Council recognised that sports can play a positive role in community well-being and social regeneration. That sports can contribute to the physical (environmental), social and economic regeneration of neighbourhoods is also stressed in a national document on the 'value of sport' (Sport England, 1999). This document aims to demonstrate the *tangible* benefits of sports to individuals and communities, in order to justify its share of the limited public resources. Manchester is one of the frontrunners in translating this document to regional policies, making grateful use of National Lottery resources. Between 1998–2001, the city received £130m from these resources, from which £90m have been reserved for the City of Manchester Stadium, £22m for the swimming pool complex and £14m for other sports venues (an indoor tennis centre and the National Squash Centre).

For pro-active development, 13 focus sports have been identified. Each sport has its own office. Manchester Leisure aims to improve access to sports opportunities throughout the city by linking schools and clubs, and to raise the image of Manchester as a city of sports on both the domestic and the world stage, helping ongoing and sustainable inward investment and success for its sporting infrastructure. Furthermore, the city wants to ensure that the new facilities for the Games create long-term benefits for the citizens of Manchester. Through Sports Development partnerships these facilities will support the city-wide development plans for each focus sport towards a structured approach to the delivery of coaching programmes, officials and coach education, sports science and international events. Manchester Leisure has created partnerships with national, regional and local sport-related organisations (such as Sport England, Manchester City Football Club, Lawn Tennis Association, etc.). The City has tried to integrate the sports strategy in its revitalisation strategy by linking sports organisations to other organisations such as schools and companies. The revitalisation policies are coordinated by seven area-specific offices in 15 wards. In these offices, civil servants from various departments work together. The area-specific offices cooperate with the sport-specific offices. In that way, links between clubs, schools and companies can be created, possibly making use of national employment programmes (New Deals).

To maximise the regeneration benefits in East Manchester, the MetroLink system will be extended to Sportcity, and the missing link in the ringroad around the city will be realised. Sportcity, which comprises not only sports facilities but also business units and dwellings, can be a catalyst for the development of East Manchester into a new economic growth pole. Its realisation will enhance the quality of the living and location environment and the image of the entire area, especially as the accessibility of the area will be improved as well. This approach fits well into the new policies of the City Council, aimed as they are at stimulating the development of employment in the (deprived) neighbourhoods instead of stimulating commuting to the city centre with its negative effects on accessibility and the quality of the living environment. One positive effect of the creation of Sportcity is the increasing attractiveness of schools in East Manchester. Because schools are linked to the facilities, they can offer a high-quality sports programme.

Furthermore, the city of Manchester aims to use the Commonwealth Games for education initiatives, on the assumption that learning through sport can: 1) play a major part in the city's regeneration strategy and the enhancement of its national and international profile; and 2) provide a positive focus for people's lives. It is a vital element of the City Council's social regeneration policies, particularly for children and young people. Manchester Leisure and Manchester Education Services have developed a number of projects to maximise the social benefits of the event, including the development of teaching material, programmes that stimulate participation in sport and a sports research and information centre. The projects will be co-financed by private sponsors. The potential benefits for these sponsors are outlined in the brochure 'Towards the Commonwealth Games' (Manchester Leisure and Manchester Education Services, 1999):

- all administrative documentation relating to the scheme being supported by the sponsor will be given recognition, for instance in letter headings, posters, certificates, etc.;
- there will be a comprehensive programme of media exposure opportunities arranged, including photocalls at events and tournaments, presentation of trophies and press releases on 'good news' items relating to individuals, clubs and schools participating in the programme;
- the sponsor will be credited on the individual project they are supporting;
- the sponsor's equipment will be used wherever possible on the project they are supporting;
- sponsor posters will be displayed prominently at the Sports Development

office and all venues used in connection with the programme being supported.

Despite all the initiatives to make the Commonwealth Games 'the inclusive games' (illustrated by the slogan 'count yourself in'), some negative developments with regard to that objective cannot be ignored. For some venues, such as the Velodrome, the conclusion must be that the interaction with its direct environment is too limited. Moreover, the social support for the Commonwealth Games is not 100 per cent. Many citizens criticise the promised benefits of the event, although scepticism may disappear as the Games draw near. One way to gain social support is to build sports complexes that combine top-class sports facilities with leisure (neighbourhood-oriented) facilities.

Sports-related Industries

Another initiative that is in some respect related to the event is a study of sports-related industries. The aim of this study is to develop a targeted strategy to attract sports and related industries to Manchester and the northwest region and to strengthen the economic base of the area using the 2002 Manchester Commonwealth Games as a catalyst (Strategic Leisure, Chesterton and Sheffield Hallam University, 1999). In particular, the study seeks: 1) to arrive at a definition of sports-related industries (a sector that to date has been poorly defined and imperfectly understood – and consequently has a low profile in terms of external perception and economic potential); 2) to audit the industry in Manchester and the northwest; 3) to identify barriers to the industry's growth in the northwest; 4) to develop an action plan, responding to local unique selling points and needs to encourage increased investment; and 5) to identify the key partners able to contribute to its achievement.

According to this study, the supply side of the sports market is a mix of three types of provider: the public sector, the voluntary sector and the commercial sector. However, the researchers have concentrated on the definition of the *commercial* sports sector, since in their opinion it is the lead sector in any strategy based on the sports-related industry, which is mainly due to the substantial economic benefits in terms of the sector's contribution to the generation of employment, the value of consumer expenditure, the investment opportunities and the 'footloose' nature of the industries within the sector. Table 4.3 shows the structure of the commercial sports-related industry.

From the study the conclusion is warranted that the Manchester region has a competitive advantage over every other UK region in the combination

Table 4.3   The structure of the commercial sports-related industry

Sports goods	Manufacturing	Sports equipment
		Sports clothing
		Sports footwear
		Alternative sports-related manufacturing
	Distribution	Warehousing
Sports services	Spectator	Professional team sports
		Stadiums/facilities
		Venue operators
	Commercial leisure	Health and fitness
		Golf
		Other private clubs
	Business services	Design
		Sport agents
		Consultancy
	Sports administration	Governing bodies of sport
	Sports science	Physiology
		Sports engineering
		Psychology
		Bio mechanics
	Media	Regional/local TV
		Regional/local radio
		Publishing

Source: *Sports-related Industry* (Strategic Leisure and others, 1999).

of sports-related industries. The region is the home base of more football clubs than any other region (including Manchester United and Liverpool). Out of the 92 clubs in the Premier League and the Football League, 21 per cent are in the northwest. The Manchester region has three of the industry leaders in sports clothing, equipment and footwear: Umbro, Reebok and Adidas. It also houses other major suppliers, such as Ellesse, which is part of the Pentland Group. That is the largest concentration of economic activity in these industries of any region in the UK, which can be partly explained by Manchester's historical status as textile city. The estimated contribution of the sports-related industries to the regional economy (3 per cent of GDP) is close to twice the national average. Another advantage of the Manchester region is the concentration of (sports-related) media, such as BBC Sports and MUTV (Manchester United TeleVision). In view of all these competitive advantages, the researchers conclude that the Manchester region has the best opportunity of any region in the UK for the successful implementation of a sports-industry strategy.

Despite that competitive advantage, the sports-related industry lacks identity and self-awareness. The industry leaders are not interested in cooperation because they are competitors. Managers of small companies are hard to activate, because they have many other things to do. In the final report, Strategic Leisure, Chesterton and Sheffield Hallam University recommend the industry to raise its self-awareness by (among other actions) hosting a conference to 'launch' the industry and its role within the region, by developing a forum for the sports-related industry, bringing together companies from across the region and industry as a whole, and by establishing an endowed university Chair or Centre within sports engineering and/or sports science, to confirm the pre-eminence of local higher education institutes within these fields, and to facilitate access by local growing businesses. These initiatives could stimulate a vision on the development of the sport-related industry. The Commonwealth Games can be used as an instrument to increase the self-awareness among businesses in this industry.

The Special Relationship between the City and the Club

Manchester is known all over the world as the home base of Manchester United, one of the most successful and valuable football clubs around the globe. In the 1998/99 season, United was responsible for one-fifth of the total turnover of the English Premier League. Moreover, it was the most successful club in Europe, winning the Champions League, the FA cup, the World Cup and the Premier League. It is listed as the richest club in the world, with an annual turnover of £100 million. The club's matches attract as many spectators as its stadium – Old Trafford – can accommodate (61,250). Recently, its capacity has been expanded by the construction of new stands.

However, few people are aware that Old Trafford is actually located just outside the official city borders, in the borough of Trafford. Consequently, there are no or few formal relations between the city and the club that seems so important for its image. As a consequence, United plays only a limited role in Manchester's marketing activities. Although the club managers and players show willingness to cooperate in promotional activities, the relationship between club and city is not as well developed as one might expect. While the club does participate in some of the City Council's overseas activities, it is not involved in local revitalisation policies. Concerning city promotion, the club plays a part in promotional videos and brochures and a closer working relationship is developing with Marketing Manchester. Moreover, the stadium is an important tourist attraction, thanks to its new museum (opened in 1998), which gives visitors an opportunity to take a tour of the stadium.

The second football club of Manchester is Manchester City, which plays in the First Division (2001) and has good chances of promotion to the Premier League. On average, the team draws 30,000 fans to each home match. After the Commonwealth Games, the club will move to the new stadium in Sportcity (the City of Manchester Stadium). It is not yet clear what will happen to the old stadium. City's turnover is approximately £10 million, but this figure stands to increase if the club enters the Premier League. The relationship between the City Council and Manchester City FC is clearly closer than that between the City Council and Manchester United. The construction of the new stadium is being financed by the city. There is an agreement between the city and the club concerning the rental of this stadium in which the city calls upon the social responsibility of the club regarding, for instance, access for all income groups (fair ticket prices). Such agreements are needed because the city is using Lottery funds to finance the construction of a facility that will have a public function for only two weeks (during the Commonwealth Games) and will be handed over to a private company (the club) afterwards.

Conclusions

The case of Manchester demonstrates that bidding for events is as such an instrument of city marketing. The two Olympic bids have resulted in the construction of venues (to show the seriousness of the bid) and contributed to the general promotion of the city (since a bidding process generates much media attention). At the bidding stage the city of Manchester already recognised the potential benefits of the event in terms of employment, urban attractiveness and general city promotion. That vision stimulated cooperativeness among the public and private actors in the region and created societal as well as political support on the local and national levels. The Olympic experience has improved the relation between the city and the national government.

The Commonwealth Games in 2002 have taken over the catalyst function from the Olympic Games. The event is regarded as an instrument to create urban attractiveness, to fight social exclusion, to promote the city and to invest in high-quality sports infrastructure. The city and other relevant actors have emphasised the need to secure maximum long-term benefits. Manchester is aware that to reach that objective is very much an organisational challenge, including the need to integrate the visions and strategies of public and private actors. Many policies in the sphere of education, sports, culture, social revitalisation and promotion are related to the event. Strategic networks have been intensified, not only with internal actors (within the city) but also with key players on the national level (such as sports associations and the national government). The creation of a vision and strategy to optimise the positive effects of sports in general and the Commonwealth Games in particular has given Manchester a competitive advantage in attracting financial resources from the national government (by means of the National Lottery). The event is also considered a stimulus for the socioeconomic revitalisation of one specific part of the town: East Manchester. Manchester uses the Commonwealth Games as a lever to upgrade the quality of life, the supply of (sports) facilities and the accessibility of the area, as well as to encourage social inclusion.

Hence, the city has identified all possible positive interactions between sports and city marketing arising from the Commonwealth Games. Policies are directed at specific internal target groups (for instance: disadvantaged and Commonwealth-originating communities) and specific external target groups (for instance, companies in Commonwealth countries). The tactical and operational marketing activities are the responsibility of the relevant municipal departments in the fields of sports, culture, education, promotion, social regeneration and economic development. Furthermore, the Commonwealth Games have been identified as a momentum to increase the awareness of producers in the sports industry. The event may result in more interaction between the actors in the sports cluster, strengthening the competitiveness of Manchester and attracting more high-grade activities to the region. The role of universities and research centres – sources of knowledge – may not be underestimated in that respect. To optimise the strength of the sports cluster, the relevant actors need to develop an integral vision and strategy for the development of the sports-related industry.

Apart from the event, the city aims to integrate sports (in general) in revitalisation strategies and city-marketing strategies. To achieve synergies, links between sports associations and other actors in the city are stimulated. However, the relation between the city and its most important trump card in the field of sports does not seem well developed. The location of the club just outside the municipal borders is a complicating factor in that respect. To some extent, Manchester United is used for brand marketing and marketing directed at external target groups (in promotional videos and by trade missions).

However, the city does not use the full potential of the club, neither for internal city marketing (for instance related to social revitalisation projects) nor for external city marketing.

The relationship between the city and Manchester City – located within the municipal borders – is quite good. The agreement between the city and the club, by which the club may rent the new venue in exchange for commitment to urban policies, has created benefits for both actors. To realise similar benefits between United and the City of Manchester, a more regional approach – together with the borough of Trafford – may be advisable, on the assumption that Manchester United is an essential part of the *regional* sports product. Another challenge is to make the club's management more aware of its ties with the entire region.

Notes

1   Source: www.stormtv.com.
2   Source: Economic Benefits Initiative brochure (MIDAS).

Chapter Five

Rotterdam

Introduction

Rotterdam is actively promoting the city as the capital of sports, a title it claims on the strength of several trump cards. First, Rotterdam is host to several annual sports events with an international air, such as the Generale Bank Marathon Rotterdam (among the fastest in the world), the ABN AMRO World Tennis Tournament (the world's greatest indoor tennis event) and the 'Concours Hippique International Officièl' (CHIO). Secondly, the city hosted five matches – including the finals – of the European Champions' Football 2000 (EURO 2000), organised jointly by the Dutch and Belgian football associations. Thirdly, Rotterdam is the home town of Feyenoord, one of the most successful Dutch football clubs, possessing the European and World Cups in its trophy cabinet.

In this chapter we will concentrate on Rotterdam's efforts to use sports as a lever for promotion and socioeconomic revitalisation. After a short city profile, we will discuss the initiatives related to EURO 2000. The next section deals with the relationships between the city and the (football) clubs within its borders, and is followed by an analysis of the interaction among city-marketing organisations in Rotterdam. The final section draws some conclusions.

Profile

With 590,000 inhabitants, Rotterdam is the second largest city of the Netherlands. From the end of the nineteenth century until well into the 1960s, Rotterdam's economy expanded rapidly, thanks largely to the prosperous growth of harbour and harbour-related activities. From the mid-1960s, industrial development stagnated. Shipbuilding and ship repair, once flourishing sectors, experienced problems, while tertiarisation proceeded slowly. As people began to set more store by well-being than by further economic growth, and thus became aware of their environment, Rotterdam acquired a bad reputation as far as the quality of life was concerned. The Rotterdam economy, vulnerable as it was, suffered heavily under the economic recession of the early 1980s.

With progressive tertiarisation, the employment structure of the city has changed throughout the years. The share of industry decreased, while business and consumer services gained ground. But the city still possesses the world's largest port and its image is still mainly harbour-related. By its policy, the city of Rotterdam tries not only to diversify the economy but also to enrich the town's image with new elements such as culture and sports. Some sectors that have been identified as growing are audiovisual services, media, recycling, and printing and publishing (RCDC, 1998). The Rotterdam economy is now less dependent on the port than it used to be. In terms of promotion, the city, together with Porto (Portugal), succeeded in being appointed Cultural Capital of Europe in 2001. Rotterdam is relatively well endowed with cultural facilities; indeed, its annual film festival has grown into one of the largest in Europe. Moreover, the city is admired for its modern architecture, with the Erasmus Bridge as figurehead.

Rotterdam's strategy (reflected in the City Council's latest programme) is based on a vision which testifies to Rotterdam's ambition to take its place as a service city and regional core of culture, entertainment, shopping and tourism, and at the same time become more attractive to entrepreneurs, residents and visitors (Van der Vegt, 1998). With its latest programme (see Table 5.1), the City Council aims to turn Rotterdam into a strong city with pleasant neighbourhoods for committed citizens. Unfortunately, some people in the city (concentrated in a few neighbourhoods) are not benefiting from the recent economic growth. To tackle that problem, several programmes to fight poverty and social segregation have been introduced. Another policy theme is the development of urban tourism. Although Rotterdam has a lot to offer, tourists seem to avoid the city. To remedy that, the sub-programme 'Attractive City' draws attention to coming events (in particular those related to Cultural Capital 2001), and is planning ways and means to improve the connections between zones likely to attract visitors. Especially people who live far away still tend to associate Rotterdam mostly with its port. The city should therefore do its utmost to upgrade its products (hotels, cafés, restaurants, events and attractions) and communicate its strengths (RCDC, 1998).

EURO 2000

The European Football Championships 2000

In June 2000, Belgium and the Netherlands were in the spotlight of the

Table 5.1　Themes of the council programme 'Met raad en daad'

Theme 1: Strong city
Economy and work: more work and a sound urban economy
Sustainable city: Rotterdam as strong city with future value
Space for Rotterdam people: Spatial Plan Rotterdam 2010 and Attractive City

Theme 2: Valuable neighbourhoods
Neighbourhood approach: major-cities policy and social activation proceed in unison
Clean and in good repair: Rotterdam should be appreciated as a clean city
Safe: more safety in Rotterdam

Theme 3: Concerned citizens
Fight against social isolation and poverty: a '100 per cent poverty policy'
Growing up in Rotterdam: the youth policy, including education

Theme 4: Enterprising government
Government 2000+: friendly approach and citizens' participation

Source: *Met raad en daad; Council's programme, 1998–2002* (City of Rotterdam, 1998) (city
　　council's action programme).

international media because of the European Football Championships 2000
(EURO 2000), the world's third-largest sports event, expected to draw some
1.5 billion television viewers around the globe. For the first time in history,
one of the three global mega-events (Olympic Games, World Football
Championships and European Football Championships) was organised by
two countries. The tournament consisted of 31 matches between 16 countries,
played in eight cities. The opening match was on 10 June in the Koning
Boudewijn Stadium in Brussels. The final was played on 2 July in the
Feyenoord Stadium ('de Kuip') in Rotterdam. The six other hosting towns
were Amsterdam, Arnhem and Eindhoven (the Netherlands), and Bruges,
Charleroi and Liège (Belgium).

The anticipated economic spin-offs of EURO 2000 were high. The
Netherlands Economic Institute (NEI) expected that 1.2 million people would
visit Belgium and the Netherlands to watch the matches. Obviously, these
visitors spent money during their visit, thus generating additional income for
hotels, restaurants, shops and other consumer services. Rotterdam hosted five
matches, including three group matches (Spain-Norway, Denmark-Netherlands
and Portugal-Germany), one of the quarter finals and the finals.

The Master Plan

As the 'city of the Final', Rotterdam sought to distinguish itself among the host cities as an international, exciting and hospitable city by staging other events before and after the five matches. These events were intended to help to prevent problems with supporters (crowd control). One great challenge was to organise an event for the local citizens. The City Council reserved more than €1.8 million[1] for all these activities, on condition that private actors contributed the same amount (Project Bureau Rotterdam EK 2000, 1999).

Rotterdam recognised the need for cooperation between several actors to make EURO 2000 a success for the city and its citizens. The EURO 2000 Steering Committee, chaired by the alderman for sports, was responsible for the organisation of activities. Represented on the Committee were the Rotterdam Leisure Department, Rotterdam Topsport, Rotterdam Festivals, Public Works, Feijenoord Stadium, Ahoy' Rotterdam, the Rotterdam City Development Corporation, the Rotterdam Chamber of Commerce, the Rotterdam Tourist Office, the Rotterdam Government Department, the Rotterdam Rijnmond Police, the Public Prosecutor and Rotterdam Marketing. Out of that steering committee, a core group – executive committee – was formed, in which Feijenoord Stadium, Rotterdam Festivals, Rotterdam Topsport, Public Works, Ahoy' Rotterdam and the Rotterdam Leisure Department were represented. This executive committee guided the Project Bureau Rotterdam EK[2] 2000. The budget was controlled by the Rotterdam Leisure Department, which appointed an external project leader and two assistants to draw up and implement a master plan. The city deliberately opted for a small project bureau, assuming that it could activate other public and private actors by relating their core business to EURO 2000.

The master plan reflected the city of Rotterdam's ambitions for its function as Finals City of EURO 2000. This ambition was translated into three interrelated objectives:

1 to ensure that the matches were properly organised (and problems avoided);
2 to optimise promotion and publicity for Rotterdam;
3 to optimise the economic spin-off in terms of income and (mainly temporary) employment.

To reach these goals, seven sub-projects were defined:

1 events and leisure pursuits;

2   communication and promotion;
3   overnight stays and tourism;
4   sponsoring and business;
5   logistics and transport;
6   order and security;
7   EURO 2000 Service Bureau.

Events and Leisure Pursuits

The purpose of the sub-project 'events and leisure pursuits' was to mount events and other activities to entertain citizens and visitors and stimulate EURO-2000 awareness. In the 12 months before the tournament, several events were planned to create societal support. Three types of event can be distinguished: crowd-control events, public events and neighbourhood events. High-risk groups of supporters were the target of crowd-control events, which were intended to take care of supporters and separate them. For a high-risk match, the organising bureau determined the type and location of activities, in cooperation with the police. Public events addressed first of all the Rotterdam population, but also welcomed visiting supporters. These events, which were concentrated in two areas on either side of the river Nieuwe Maas, were intended to promote Rotterdam as an attractive and hospitable city. Neighbourhood events were oriented to the local residents. They fitted in well with the social revitalisation strategy of the city, including the so-called 'Opzoomer Mee' projects, aimed at raising the quality of the living environment at street level. Activities were in the nature of sports, games, art and culture. Thus, EURO 2000 was inteded to help stimulate social cohesion and create a pleasant atmosphere.

The three types of event were integrated in the communication strategy. The invention and initiation of public and neighbourhood events was left to private actors. An independent committee judged any plans by specific criteria and preconditions and acted as advisor to the core group. The members of the core group (advised by the steering group) decided whether or not the plans deserved support. Crowd control events were developed by the project bureau.

Communication and Promotion

The objective of the sub-project 'communication and promotion' was to promote Rotterdam – in its capacity as Finals City of EURO 2000 – as an attractive, hospitable, colourful and modern city with many opportunities.

One element of this sub-project was the design of a logo and 'city dressing'. Internally, the city administration aimed to create a EURO 2000 feeling among the Rotterdam population and the project bureau cooperated with the local media. Target groups of the external communication strategy were (potential) visitors, officials and the press. A brochure on Rotterdam was distributed by mailings, exhibitions, conferences and sales missions abroad of the Tourist Office and Rotterdam Marketing. For these activities, the project bureau worked together with Rotterdam Marketing. Thanks to a joint initiative of all the host cities of EURO 2000, a collective brochure was also released.

Volunteers from the Rotterdam community served as hosts and hostesses, providing information about attractions, activities, transport, accommodation, etc. Initiatives related to hospitality were managed by a partnership of the project bureau, the organisation Rotterdam Cultural Capital and the Rotterdam City Development Corporation, in cooperation with 'Opzoomer Mee' and the City Council's Department for External Relations.

Overnight Stays and Tourism

The objective of the sub-project 'overnight stays and tourism' was to attract visitors and entice them to stay longer and spend more. To reach the target group, information packets were sent to foreign tour operators. The organisation of day trips to other host cities (which in turn organise day trips to Rotterdam) is one product developed in that framework; another idea is the Rotterdam Access card (which offers access to all attractions and museums in Rotterdam as well as to public transport). The introduction of the Rotterdam Access Card was related to the RCDC project 'Internal Accessibility'. Furthermore, a reservation system for hotels in the Rotterdam region was installed. EURO 2000 clearly stimulated the creation of innovative tourist products.

Sponsoring and Business

The objective of the sub-project 'sponsoring and business' was to stimulate business companies (in the Rotterdam region) to participate with financial, material or other contributions. In an attempt to double the municipal budget of 4m guilders, the project bureau initiated a business-partner programme through which companies could participate in the events. Three sponsoring packages were composed. The Champions Club package was meant for (inter)national companies who attached much (financial) value to EURO 2000. It included hospitality packages, the use of the Rotterdam EURO 2000 logo, branch-exclusiveness and the 'right of first refusal' for five events. Official sponsors and suppliers of EURO 2000 automatically received this package. The target group of the Runners-up Club package comprised the national and regional medium-sized businesses. This package gave participants the right to use the logo and to promote the company at three events. The Qualifier Club package offered companies in the Rotterdam region the opportunity to participate in EURO 2000 against payment of a relatively small amount. For this sub-project, the project bureau cooperated with ISL and other sports-marketing organisations.

Logistics and Transport

This sub-project aimed to channel the visitor flows to and in Rotterdam. To that end, a regional mobility plan was designed to create policy conditions for the transportation of visitors who arrived in Rotterdam with the aim that visitors should be able to reach and leave the Feijenoord Stadium without causing inconvenience around the stadium. Specific plans were drawn up in cooperation with public transport companies. One of the ideas suggested in this sub-project was 'remote parking', which means that visitors parked their car outside the city to continue their trip by public transport (Park and Ride). Regrettably, this was only a temporary measure. Another point of this sub-project was to stimulate the use of (electronic) signposting in and around Rotterdam.

Order and Security

The threat of hooligans disturbing EURO 2000 was a major concern in the national media and to national politicians. It was inspired by the tragic incidents during the World Football Championships in 1998 and some other recent problems with club supporters in the Netherlands. One of the worst incidents followed upon FC Feyenoord's celebration of its national championship in 1999, when hooligans from all over the country organised a riot. These incidents gave rise to a strategic reorientation of the police's approach.

Negative media attention can frustrate all initiatives to use sports for city-marketing objectives, since it reduces support among companies (sponsors) and the local population (internal target groups of events and matches). For EURO 2000 the city preferred to adopt a public-friendly approach. Crowd control events were initiated to take care of supporters. The police were assisted by the Rotterdam hosts and hostesses, who provided visitors with the information they needed. A number of information posts were erected to inform supporters where the various activities were staged. Special Patrol Groups interfered only in case of disturbances. The police were prepared to take measures to restrain the consumption of drugs and alcohol.

Service Bureau EURO 2000

The Service Bureau EURO 2000 functioned as service point for the organising committee and affiliated companies and as intermediary between the EURO 2000 Foundation and the Rotterdam business companies. It aimed to attract core business activities of companies affiliated with EURO 2000 in the fields of communication, promotion, accommodation, media, sponsoring and city dressing. The Service Bureau's main function was to mediate between local business and ISL in matters of sponsoring.

Lessons from EURO 2000

The efforts of the city to use EURO 2000 as a lever for city marketing have inspired knowledge among the actors in the network. Although at the time of writing the event was still at a preparatory stage, at least two lessons can already be drawn from the experiences.

The project bureau managed to triple the municipal budget of approximately €1.8 million, thus easily exceeding the target, which was to double it. Most companies were mainly interested in exposure-generating activities directed at external target groups. Sponsors for activities focused on internal target groups (neighbourhood events) were harder to find. Despite this success, it is felt that the optimal result was not achieved, due (among other things) to a lack of *strategic* commitment among the actors, the relatively small budget and the lack of formal competencies.

Second, Rotterdam had to establish strategic relations with actors outside the city, such as the UEFA, the Foundation EURO 2000, ISL (sponsoring) and EBU (broadcasting rights). Agreements with ISL imposed the limits on city dressing, sponsoring and the use of logos, since ISL represents the interests of the sponsors of UEFA. It is important to reach an agreement with that actor at an early stage. The Project Bureau Rotterdam EK 2000 succeeded – albeit rather late in the day – in getting city dressing removed from the contract with ISL. The responsibility for city dressing *outside* the stadium area was transferred to the city on condition that the sponsors of UEFA received first right of refusal. This change of policy left more opportunities to use the event for city marketing and could be a model for the future.

Sports Clubs and the City

The city of Rotterdam wants to use EURO 2000 as a lever for promoting Rotterdam as a sports city, on the assumption that the city has a competitive advantage in the field of sports, as is evident from its comprehensive sports infrastructure (see Table 5.2).

Table 5.2   Sports facilities in Rotterdam

Athletics tracks	3	School sports fields	15
Gyms	225	Sports halls	30
Handball fields	9	Sports Palace ('Rotterdam Ahoy')	1
Hockey fields	25	Tennis courts	174
Baseball/softball fields	9	Top-sports centre	1
Marinas	32	Football fields	239
Boule tracks	63	Football stadiums*	3
Korfball fields	33	Swimming pools (indoor)	12
Skateboard tracks	2	Swimming pools (outdoor)	2
Rugby fields	2		

*   Home bases of professional football clubs.

Source: Inventory of Sports Facilities in Rotterdam, 16 March 1999 (Rotterdam Leisure Department, 1999).

Rotterdam, Football City

Rotterdam is the home base of three professional football clubs (Feyenoord, Sparta and Excelsior), which is more than any other city in the Netherlands. Feyenoord belongs to the traditional top three of the Dutch 'Eredivisie' (Premier League), together with Ajax Amsterdam and PSV Eindhoven, and attracts supporters from all over the country. In 1998 the average number of spectators exceeded 28,000. Sparta, playing in the Eredivisie as well, is one of the oldest football clubs in the Netherlands. Its home base is the recently-renovated ENECO Stadium. At the time of writing, Excelsior plays in the Eerste Divisie (First Division). This team plays its home matches in the Akai Stadium, also recently renovated. Excelsior has a close relationship with Feyenoord, which manifests itself in the exchange of players.

Although Feyenoord is a foundation and has no plans to change its legal form, the club's management is involved in a process of professionalisation. Since 1995, the club has paid more attention than before to non-football activities. Consequently, the management team includes financial and commercial managers as well as technical (football) managers. The club is increasingly aware of its relationship with its direct environment, as is illustrated by an initiative called Feyenoord Youth Project. The aim of this project is to acquaint children of disadvantaged neighbourhoods with sportsmanship and companionship. In that fashion, the club is investing not only in the social-economic strength of the neighbourhoods but also in its future supporters.

Feyenoord plays its home matches in the Feijenoord Stadium,[3] located on the south bank of the river Nieuwe Maas. At present, the capacity of 'de Kuip' – as the stadium is popularly called – is 51,177. In 1995, the stadium was renovated and turned into a multifunctional venue for sports, music concerts and conferences. One part of the renovation was the construction of business seats (400 at present) and business units (40). Another part was the construction of a building next to the stadium ('het Maasgebouw'), which accommodates a conference and event centre including seven conference rooms and a brasserie (pub). The renovation was financed by public and private actors. The city contributed an amount of more than €13.6 million and the Rotterdam Municipal Port Management provided a subsidy of over €4.5 million. The remaining €34 million were put up by private companies. Thanks to the renovation and the new activities, the number of employees of Feijenoord Stadium (a public limited company) has increased from 15 to about 65. The company also renders catering and ticket services.

'De Kuip' belongs to the elite group of European stadiums that have been awarded five stars by the European Football Association (UEFA) on several quality criteria. Consequently, the stadium met all the requirements to host the finals of the European football competitions. On 2 July, the stadium was the scene of the EURO 2000 finals. The Foundation for Professional Football 'Feyenoord' is the most important tenant of the stadium, accounting for half of the turnover. National football team matches and pop concerts are responsible for one-quarter of the turnover; the remaining quarter is derived from conferences. Feyenoord owns 29 of the 40 business units and 250 of the 400 business seats; the other units and seats are owned by the stadium. Obviously, the stadium and the club have common interests.

The renovated Sparta stadium (ENECO-stadium) is located in Delfshaven, a borough in the western part of the city that suffers from serious socioeconomic problems. The renovation has been heavily subsidised by the city of Rotterdam, as part of the revitalisation policy for this town quarter. The stadium is also used by the borough Delfshaven for various activities.

More Teams and a New Topsport Centre

Apart from the three professional football clubs, Rotterdam boasts several other sports clubs that play on the national level. Rotterdam's baseball club – Neptunus – belongs in the nation's best clubs and recently moved to a new stadium. Another sports club enjoying national success is the table tennis club FVT/Visser. Furthermore, Gunco Rotterdam Basketball and Volleyball Club Nesselande can be regarded as leading clubs. They play their matches in the recently opened Topsport Centre adjoining the Feijenoord Stadium. On average these teams draw some 600 spectators, with peaks of 2,000 or 3,000 supporters at important matches.

The Topsport Centre has been entirely financed by the city of Rotterdam (the construction costs have been estimated at almost €8.7 million). In 1990, the City Council observed a missing link in the sports infrastructure: what Rotterdam needed was a new venue with a smaller capacity than Ahoy' and the Feijenoord Stadium, with facilities for spectators, to stage indoor sports events. In 1998, the City Council agreed to give the Rotterdam Leisure Department credit to construct such a facility near the Feijenoord Stadium. The Feijenoord Stadium and Ahoy' Rotterdam were asked to participate in the construction, but they both refused. The official name of the venue is 'Topsportcentrum Stad Rotterdam Verzekeringen', named after the main sponsor, a Rotterdam-based insurance company. Besides basketball and volleyball matches, the 2,500-seater venue will host judo and boxing events as well as non-sport events such as music concerts attracting audiences too small for Ahoy'. The venue officially opened its doors in January 2000.

The Topsport Centre is a joint project of four Rotterdam institutions. The city (the Rotterdam Leisure Department) owns the facility, Ahoy' is the operator, Feijenoord Stadium is responsible for catering services and Rotterdam Topsport sets itself to attract national and international events. To optimise the occupancy rate, Rotterdam Topsport has concluded agreements with national sports associations. In consequence, the National Basketball Association uses the venue as home base for the national teams. Moreover, the centre accommodates the national judo and boxing championships. To optimise the exploitation and the social returns, the venue also hosts leisure sports events (matches) and training courses. The four institutions have recognised the need to make the Topsport Centre complementary to other facilities, such as Ahoy' and Feijenoord Stadium. The new venue can be used as a test field for new events. When an event appears successful (and the potential number of visitors exceeds the capacity of the complex), it can be moved to Ahoy'. While Ahoy' specialises in commercial sports events, the Topsport Centre can (also) accommodate non-commercial or commercially less attractive events. Moreover, the three venues (Ahoy', Feijenoord Stadium and the Topsport Centre) can form a complex during mega-events. That was the case during EURO 2000, when the Topsport Centre functioned as press centre.

Rotterdam, Event City

For its status as a city of sports, Rotterdam relies in part on the annual and incidental events that take place there. Since 1981, Rotterdam has made a success of one of the world's fastest marathons: the Generale Bank Marathon Rotterdam. From 200 participating athletes in the first marathon, the number has increased to 10,000. The organiser of this event is the Foundation Rotterdam Marathon, supported by the city. Another prestigious event is the 'Concours Hippique International Officièl' (CHIO) in the 'Kralingse Bos', one of the city's parks. Rotterdam is also proud of its annual ABN AMRO World Tennis Tournament, organised by Ahoy' Rotterdam in cooperation with a sports-marketing company (Spo Mark BV). The tournament takes place in February and attracts more spectators than any other indoor tennis event. Among the spectators are many business managers and their clients who occupy business seats and hospitality units to combine sports with business. The success of the tournament cannot be isolated from its location: Ahoy' Rotterdam.

Ahoy' Rotterdam is a sports and event centre located in the southern part of Rotterdam. Its construction was entirely financed from municipal budgets. After its opening in 1970, the venue has developed from a sports centre with adjoining exhibition and fair accommodation into a unique multifunctional complex. In the course of time, Ahoy' has been privatised. Today, the venue is owned by a public limited company: Ahoy' Rotterdam NV. The complex consists of a sports venue ('Sportpaleis', Sports Palace) with a capacity of 10,500 seats, and six exhibition halls covering a total area of 30,000 m^2. Recently, Ahoy' was expanded by a new hall measuring 10,000 m^2 and a prestigious entrance hall including ticket sales counters, conference rooms, two restaurants and kiosks. Ahoy' accommodates fairs, conferences, seminars, company parties, sports events and exhibitions. Increasingly, several activities are combined, as is illustrated by the ABN AMRO World Tennis Tournament. The success of this event is due to the unique combination of a sports venue and exhibition halls, offering a lot of space for business-to-business activities.

Business companies seem to appreciate these facilities as the great demand for business boxes and hospitality packages demonstrates. However, the number of sports events is relatively small. Apart from the tennis tournament, Ahoy' accommodates the annual Holland Basketball Week and the National Indoor Korfball Championships. Ahoy' was also the scene of the World League Volleyball Finals in 1996.

Sports and City Marketing

Rotterdam Topsport

In 1991, the city of Rotterdam took the initiative to establish the Foundation Rotterdam Topsport. Its mission is to enhance the conditions for top-class sports in Rotterdam. To that end, Rotterdam Topsport cooperates with the city of Rotterdam, national and international sports associations, business companies, educational institutions and the media. The budget of Rotterdam Topsport is composed of municipal contributions and revenues from sponsorships. Areas of special attention are events and infrastructure.

In the way of events, Rotterdam Topsport attracts, accompanies and supports new sports events in Rotterdam, such as a Davis Cup Tennis match in 1994, the World League Volleyball Finals in 1996, the All Star Basketball Gala and the finals of EURO 2000. Moreover, Rotterdam Topsport supports and if necessary strengthens regular sports events, such as the Rotterdam World Port Baseball Tournament, CHIO, ABN-AMRO World Tennis Tournament and the Holland Basketball Week. Rotterdam Topsport maintains close relations with national sports associations which play an important role in attracting international events. All these activities are directed at promoting sports (participation), city promotion and economic spin-offs.

With regard to infrastructure, Rotterdam Topsport supports clubs and organisations with top-level ambitions. Support is given to youth training, efforts to reach and maintain top level, event organisation, the use of accommodation, sponsor acquisition, promotional activities, medical care, etc. Because of the growing demand for support, Rotterdam Topsport has selected a number of key sports: basketball, hockey and volleyball (spearhead sports: high potential), table tennis and baseball (attention sports: medium potential) and badminton and judo (care sports: low potential). Each key sport is related to one top club in the Rotterdam region that receives (financial) support from Rotterdam Topsport. In exchange, the Foundation can claim the help of some players for clinics in deprived neighbourhoods to stimulate sports participation. Regrettably, football – by far the most popular sport in the Netherlands – is not one of the spearhead sports. It seems difficult to involve the professional football clubs in the activities of Rotterdam Topsport. One reason is perhaps that professional football players have many commitments to the club and its sponsors. Another may be the apparent lack of commitment among the professional football clubs to the city and the citizens.

The Foundation also supports and accompanies individual sportsmen and sportswomen. Furthermore, Rotterdam Topsport has developed facilities to combine sports and education (in cooperation with schools in the region) as well as medical facilities (the Sports-medical Centre Rotterdam, part of the Foundation). Finally, Rotterdam Topsport advises event organisers and sports associations on the total supply of facilities in the Rotterdam region (including the Topsport Centre, Ahoy', Feijenoord Stadium, the ENECO Stadium, the AKAI Stadium and the Neptunus Baseball Stadium).

City Marketing

Rotterdam Topsport belongs to a group of organisations in the Rotterdam region that are engaged in city marketing. According to Van der Vegt (1998), this so-called city-marketing network consists of public and non-profit organisations on the one hand and private actors on the other. Among the private city marketeers are event organisers and venue operators. Their objective is to promote their own event or venue. The public actors can be divided into governmental institutions, cluster-specific institutions and territorial institutions. Among the governmental city-marketing institutions are the Rotterdam City Development Corporation, the Rotterdam Municipal Port Management and the Rotterdam Leisure Department.

The Rotterdam City Development Corporation is responsible, as agent of the city, for the spatial and economic development of Rotterdam, together with the Rotterdam Municipal Port Management. The Rotterdam City Development Corporation is charged with the 'dry' areas, while the Rotterdam Municipal Port Management is entrusted with the 'wet' areas. The former is engaged in developing and promoting the residential, business and leisure areas in the city and also acts as the city's developer and real-estate agent. Consequently, the Rotterdam City Development Corporation is also involved in developing the sports infrastructure of Rotterdam. The Rotterdam Municipal Port Management has been charged by the city to develop, administer and operate the port and the industrial area (owned by the city). It sponsors various sports events, such as the World Port Tournament (baseball), the CHIO and the Generale Bank Marathon Rotterdam. Moreover, the Port Management is one of the 'founding fathers' of the Feijenoord Stadium, which owes its recent renovation partly to a substantial contribution from the Rotterdam Municipal Port Management.

The Rotterdam Leisure Department is responsible for urban policies related to sports and leisure. The department operates the municipal sports facilities, organises sports events and orders the construction of new facilities. Consequently, the department helps to enhance the attractiveness of the neighbourhoods and the city as places to live and work in (Rotterdam Leisure Department, 1998). Moreover, the department functions as advisor to the City Council and other departments in matters of open-air recreation, and subsidises many institutions. The Rotterdam Leisure Department (focusing on leisure sports) and Rotterdam Topsport (concentrating on top-class sports) complement each other well. The two organisations harmonise their policies and actions to strike the perfect balance between leisure sports and top-class sports.

Rotterdam Topsport is one of the cluster-specific city-marketing institutions. Although Rotterdam Topsport works at strengthening the relationship between sports and city marketing, its primary objective is to develop the sports sector. Another sector-specific city-marketing institution is Rotterdam Festivals. This Foundation does not itself organise events, but stimulates their staging in Rotterdam by enthusing, stimulating and accompanying event organisers. The task of this Foundation is mainly the implementation of the municipal festival policy. Rotterdam Festivals communicates the strengths of Rotterdam as well as the positive developments that have changed the city's image to the world at large. The Foundation is involved in practically all events of importance in the city centre of Rotterdam (mainly open-air festivals), with the exception of sports events and the annual film festival. Rotterdam Festival has done much for the appeal of the culture and event cluster in Rotterdam as well as for the attractiveness of Rotterdam as a whole. Examples of other cluster-specific city-marketing institutions are the Cruise Foundation (which tries to put Rotterdam on the map as a cruise city), the Foundation for Industrial Tourism and the Rotterdam Congress Bureau.

The ambition of territorial city-marketing institutions is to promote a specific area larger than, smaller than or coinciding with the city of Rotterdam. To this group of institutions belongs the Tourist Information Office (VVV), a marketing and promotion organisation in the sphere of tourism and recreation that tries to attract more visitors to Rotterdam and persuade them to stay longer and spend more. Other institutions of the same kind are the Water City

Association (Vereniging Waterstad), the Port Promotion Council, the Chamber of Commerce and shopkeepers' associations.

Apparently, then, there are plenty of city-marketing initiatives in Rotterdam, but they seem to be poorly attuned to each other. Most initiatives are fragmented for lack of an integral city-marketing policy to bind all the pieces of relevant knowledge together. Nor is there a common vision or strategy to guide all these city-marketing efforts towards a common goal.

The city of Rotterdam, aware of the deficiency in its city-marketing network, has established a new organisation – Rotterdam Marketing – whose mission is to harmonise existing marketing and promotion activities and optimise their benefits. This new organisation is to design an integral marketing and promotion policy and stimulate cooperation between governmental, cluster-specific and territorial city-marketing institutions on the one hand, and private companies such as venue operators and event organisers on the other. Rotterdam Marketing has taken over the marketing and promotion activities of the Rotterdam Tourist Information Office (VVV), which will retain its front-office function. In the end, Rotterdam Marketing is to evolve into a private limited company with the city of Rotterdam as its main shareholder. To safeguard the cooperativeness of the actors in the city marketing network, Rotterdam Marketing is intended as a substitute for rather than a complement to existing marketing organisations.

Conclusions

The city of Rotterdam looks upon sports as a lever to make the city more attractive. In the last few decades, Rotterdam has continuously invested in its sports product by building venues, organising events and stimulating the practice of sports, not only to create a healthy city, but also to fight social exclusion. With respect to top-class sports, the city has gained a competitive advantage in the local, national and even international context by stimulating top-class sports, constructing and improving venues (such as Ahoy', de Kuip, the new Topsport Centre) and creating the right conditions for events (tennis, marathon, football, basketball, baseball and equestrian sports).

The city of Rotterdam was aware that the European Football Championships 2000 offered a unique opportunity for successful city marketing. From the Masterplan, most activities appeared to be directed at promotion and marketing rather than product development or (direct) investment in the city's attractiveness. Although EURO 2000 instigated new tourist products (such as a new reservation system), it was not used to construct or improve facilities or infrastructure. One explanation is that Rotterdam is already so well endowed with high-grade sports products and complementary infrastructure that additional investment was less urgent than in many other cities. Nevertheless, Rotterdam could have made more of the event as a catalyst towards cooperativeness and a stimulus to invest in the city, for instance by relating the event to social revitalisation.

Some of the activities that Rotterdam organised in the context of EURO 2000 Rotterdam were addressed at *internal* target groups. The city regarded the event as an instrument to create solidarity and a community spirit among the Rotterdam people. The Masterplan, too, referred to a project intended to raise the quality of life in the neighbourhoods and thus their attractiveness. On the other hand, EURO 2000 was also considered a city-marketing instrument oriented to *external* target groups. The city tried to attract visitors at the time of the event and to increase their length of stay as well as their expenses. However, most marketing initiatives had a short-term perspective, being focused on the event itself (and its direct spin-offs) rather than on its catalyst function (and its indirect spin-offs). The municipal efforts to use sports for city-marketing risked being thwarted by safety concerns that could reduce the effectiveness of city promotion and imaging and of any marketing efforts directed at target groups inside and outside the city.

Rotterdam could have made more of its competitive advantage in sports in general and of the unique opportunity offered by EURO 2000 in particular. For EURO 2000, too many projects were defined on an ad-hoc basis and the objectives were related mainly to the event as such. Although the project bureau was successful in reaching the financial targets, it is felt that more benefits for the city could have been achieved with a higher budget, a stronger strategic commitment among the actors involved and more formal competencies of the project bureau. In the circumstances, organising capacity seemed to be of the essence. To get the necessary (financial) support and commitment from companies and municipal departments, a vision and strategy, setting the ambitions and objectives, should have been drawn up at an earlier stage.

Nor are the organisational challenges restricted to the event as such. To optimise the synergies between a sports event and city marketing, a strong city-marketing organisation is needed. Until recently, city-marketing initiatives in Rotterdam were fragmented and unrelated. In that context, the establishment of a new city-marketing organisation (Rotterdam Marketing) is promising. It may be a first step towards an integrated sports and city-marketing policy that relates sports to tourism, culture (by strengthening the relation between EURO

2000 and Cultural Capital 2001 for instance) and the socioeconomic revitalisation of deprived neighbourhoods.

First of all, the relationship between the city and the professional football clubs needs to be further developed. Considering the popularity of football, the clubs should play a much more explicit role in the economic and social revitalisation policy of the city of Rotterdam. These football clubs do not seem convinced of their own ties with the city. Furthermore, the city should develop a medium- or long-term event strategy that regards events as catalysts for creating an attractive city. Rotterdam should start thinking about new events after EURO 2000 and the Cultural Capital of Europe 2001. In that view, the plans to organise a maritime event in Rotterdam – within three or four years – are promising.

Notes

1   All amounts in this chapter have been converted by the official Dutch guilder rate into Euros: 100 guilders are equal to €45.378.
2   'Europese Kampioenschappen' (European Championships).
3   Note the spelling difference between the name of the football club 'Feyenoord' and the stadium 'Feijenoord'. The club prefers the more international style with an upsilon ('y') instead of the typically Dutch dip thong 'ij'.

Chapter Six

Turin

Introduction

Turin owes its selection as one of the case studies in this investigation to its successful bid for the 2006 Winter Olympics, leaving (among others) the cities of Sion and Helsinki (see Chapter Three) behind. Among the people of Turin, there is a general feeling that this event can help the city to acquire a new image, of which sports and culture are major elements. Another motive for analysing Turin is the presence of one of the most successful football clubs in Europe – Juventus – within its borders. Just like Barcelona (see Chapter Two), at the time of writing Juventus is involved in an interesting discussion with the City Council. We will also pay attention to some other interesting sports-related projects. Most of the initiatives are intended to increase participation in sports ('sports for all').

The remainder of this chapter is divided into five sections. The first section contains a brief city profile. After that, the Olympic bid comes up in section two, followed by the ambitions of Juventus in section three. The fourth section deals with several initiatives to stimulate participation in sports. The fifth section concludes.

Profile

The city of Turin is located in northern Italy at a distance of 80 km from the French border and 100 km from the Swiss border, thus offering easy access to several popular ski resorts in the Italian Alps, of which Sestrière is probably the best known. Turin (with almost 1 million inhabitants, at the time of writing, Italy's third largest city) is the capital of the Turin Province (2.2 million inhabitants) and the Piemonte region (4.3 million inhabitants).

In the last 100 years, Turin's ups and downs have followed the growth and decline of one company: Fabbrica Italiana Automobili Torino, better known as FIAT. After its establishment in 1899, this company grew into one of the world's largest car producers. In the late 1970s, FIAT employed approximately 150,000 persons in the Turin region, compared with 78,000 in 1960. The growth of industrial activities resulted in an inflow of immigrants from southern Italy, which can be illustrated by the fact that the population grew from 700,000 in 1940 to 1,200,000 in 1974. In consequence of the geographical concentration of the Italian car industry in the Turin region, the restructuring of the global car industry in the 1970s and 1980s affected the Turin economy heavily. Since 1971, the reorganisation of existing plants and the relocation of activities to southern regions in Italy have resulted in a loss of 65,000 jobs in the Turin region. Despite these dynamics, FIAT and Turin still have strong mutual links. However, the relationship has shifted from manufacturing to service activities. At present, Turin functions as a major intelligence centre for car design, production and marketing in the world.

Today, the employment structure of Turin is dominated by the service sector (56 per cent). However, that is significantly less than in other cities in Italy and Europe. Some of the sectors that have developed are banking, insurance and publishing. Several companies in these sectors have their headquarters in Turin. The city's accessibility improved with the opening of a new international airport in 1993 and will improve further by its connection to the network of high-speed trains. The cultural sector is also well represented in Turin. The city was among the first in Italy to import and experiment with the film technique in the early years of the twentieth century, to realise the artistic importance of this new medium, and to make films that were shown all over Italy. Turin hosts four international film festivals, including the Festival of Youth Cinema which has achieved genuine international status over the years. Among the cultural facilities in Turin are an opera theatre, three auditoriums, 12 theatres, 53 cinemas and 26 museums. The most renowned museum is the Museum for Egyptian Art, one of the richest in the world, just behind that of Cairo.

Turin is not a popular tourist destination. In 1996, the total number of nights spent in hotels was 1,293,440 (Van den Borg, Russo and Rumi, 1999). Most hotels are oriented to business travellers: 59 of the 124 hotels are in the three- and four-star categories. The weak development of leisure tourism can be partly attributed to Turin's poor image. Especially foreign, but also Italian visitors, see Turin as an unattractive industrial centre that has not much to offer to them. Actually, Turin does have a great deal to offer, in particular to cultural tourists, in consideration of the number of cultural facilities. Moreover, the historical city centre – built by the Romans – is interesting from an architectural point of view.

As depicted in Table 6.1, the region of Turin has considerable experience in organising sports events. An annual international sports event is the Turin

Marathon. Its first occurrence was in 1919, which makes it one of the oldest marathons in the world. The international fame acquired by the Turin Marathon in the past years have made it an event that attracts a considerable number of tourists to the city and that contributes to the promotion of the city and its products. In 2001, the city hosted the World Heel-and-toe Walking Championships. Just like cycling and running, walking is an ideal sport to promote the city by television coverage, especially since the event will take place right in the city centre.

Table 6.1   Sports events in the Turin region

Alpine skiing	1997 World Championships, Sestrière; Slalom della Gazetta, Sestrière (annual)
Athletics	1997 World Cross Championships, Turin, Turin Marathon (annual)
Basketball	European Club Cup Final, Turin, 1993
Cycling	Milan-Turin (annual); Tour of Piedmont (annual)
Football	World Championships Football, 1990 (co-host)
Rowing	Gran Fondo d'Inverno, Turin (annual)
Volleyball	World League Quarter Finals, Turin and Cuneo, 1995

Source: *Torino 2006* (Associazione Torino 2006, 1998), adapted by the authors.

The 2006 Winter Olympics

In 2006, Turin will host the Olympic Winter Games, one of the biggest sports events in the world. In the competition for the event, Turin defeated Helsinki (Finland), Sion (Switzerland), Zakopane (Poland), Poprad-Taty (Slovakia) and Klagenfurt (Austria).

The bidding stage

The promotion of the candidacy was a joint venture between the city, the province and the Piemonte region. Political support was expressed by the national government, the National Olympic Committee and the national sports federations. The percentage of people in favour of the bid (societal support) amounted to 81.3 per cent in July 1998, while only 4.2 per cent were against. The motive of Turin's candidacy was the belief that the event could help the region: 1) to promote its tourist product; 2) to enter a new stage of economic development; and 3) to break out of its stereotyped image (Associazione Torino 2006, 1999).

In the proposed concept for the Winter Olympics in Turin, the snow sports (skiing, bobsleigh, luge and biathlon) will be staged in the Suse Valley and in Pragelato (more than 80 km from Turin). The region of Turin has great experience in organising alpine skiing events. The majority of the alpine events will be held in Sestrière, 75 minutes from Turin. Because the 1997 Alpine World Championships were staged here, the infrastructure is of a very high standard. The other alpine events take place in Bardonecchia (combined slalom events for men and women, snowboard) and San Sicario (Super G, women's downhill and women's combined downhill). San Sicario will also be the venue for biathlon. Cross-country skiing, Nordic combined and ski jumping will be organised in Pragelato, where a new 20,000 capacity ski-jumping stadium will be constructed. Freestyle skiing will be held at Sauze d'Oulx.

In the city of Turin itself only ice sports will be staged. Men's ice hockey will take place in two venues: a new 12,000-seater multifunctional arena to be built in Continassa close to the Olympic Village and at Espozisioni, an exhibition hall with 6,000 seats where temporary installations will be set up. The women's ice-hockey tournament will be mounted in an existing facility in Pinerolo with a spectator capacity of 3,200. The figure-skating and short-track speed skating competitions will take place at Palavela in Turin, an existing building 20 minutes from the Olympic Village, where a temporary major ice rink and spectator stands for 10,000 will be built. Speed skating will be held in a new facility (Palasport Velocita) to be built at Continassa for 10,000 spectators, in the vicinity of the ice-hockey stadium. The venue for the curling competition will be a new permanent ice facility (with a spectator capacity of 3,500) which will be erected on the site of an existing open-air rink. The Italian government has provided a US$616 million guarantee which is to cover capital investments and any additional deficit arising from the organisation of the Games.

The main Olympic Village will be located in the 'zona delle Ferriere', near the historical city centre of Turin and the stadium for the opening and closing ceremonies: the 70,000-seat Delle Alpi football stadium. This former industrial area has been assigned by the city as one of the major cores of the new post-industrial urban development. This area will be transformed from an unattractive collection of abandoned factories into a new urban district characterised by a high quality of the living environment: new dwellings in the northern part, a new railway station to the east, business and consumer services in the south, all around a large new city park. A part of the Dora river (more than 500 metres) will be uncovered (it is now covered by concrete!) and its banks restored. This so-called Spina-3 district will have good public and private transport connections to the city centre and the airport. The construction of the Olympic Village is regarded as a stimulus to launch the transformation of the area (Associazione Torino 2006, 1998b). The two other Olympic villages will be constructed in the mountains and in Lingotto (the Media Village).

Investments

The candidacy for the Olympic Games has resulted in the creation of a network of institutions that are prepared to invest in the attractiveness and promotion of the city. The event is a potential catalyst for the economic development of the entire metropolitan region. The Olympics will be an exceptional opportunity to design and build the future of Turin, providing extra impetus to the process of renovation that the city has set in motion with its strategic Torino Internazionale plan (City of Turin, 2000b).

Torino Internazionale

The starting point of the strategic plan is a long-term vision, in which the city expresses its wish to develop Turin into a European metropolis, a creative city that is producing and learning and a city that has opted for future intelligence and quality of life (City of Turin, 2000a). To realise that vision, the assignment as host of the 2006 Olympics is considered a great occasion to reveal and accelerate the required transformation.

The organisation of the Games has stimulated the relevant public and private actors to cooperate in order to be ready in time. Being aware that Turin is observed by the world and knowing that the deadline of 2006 cannot be postponed, the event will certainly ease the realisation of investments in the framework of the strategic plan. Turin has transformed the Olympics from a two-week event into a catalyst for the transformation of an entire region. Moreover, the event is considered a source of pride, optimism, cohesion and a new mentality – openness towards changes – among the population as well as among public and private actors in the region. Their cooperativeness makes it easier to strengthen the network that is needed to realise the Torino Internazionale plan.

This strategic plan will be implemented along six so-called strategic lines (City of Turin, 2000a):

1 integrating the Turin region in the international network of cities by improving external and internal accessibility;

2 reinforcing the metropolitan government;
3 strategically investing in education and research;
4 attracting new economic activities and employment (by stimulating the creative potential and providing the right conditions);
5 promoting Turin as a city of culture, tourism, commerce and sports;
6 improving the urban quality.

In relation to sports, the fifth strategic line is particularly interesting. From the strategic plan, the city looks upon sports as a vehicle for tourism, since events attract both participants and spectators. The Turin Marathon attracts both target groups, while football matches mainly attract spectators.

The Olympic Games can be used not only to promote the city nationally and internationally (the fifth strategic line), but also to realise infrastructural projects (the first strategic line) and specific facilities as well as to stimulate the diversification of the economy (the fourth strategic line). Among the projects are new sports facilities and investments aimed to improve the communication and transport infrastructure.

Transport Infrastructure

The need to invest in the transport infrastructure – to improve the internal and external accessibility – can be justified by the growing demand for mobility in the Piemonte region. Therefore, most investments are not directly related to the Olympics. Nevertheless, the deadline function of the Olympic Games is expected to accelerate the projects. One project refers to the completion of a circle of highways around the city by filling in some missing links. Another project deals with the rail connection of the airports of Turin (15 km north of the city) and Milan (Malpensa) with Turin's city centre. Furthermore, the number of railway lines crossing the city of Turin in the north-south direction will be quadrupled and the connections between Turin and the skiing areas will be enhanced by improving existing roads and the construction of a light rail system. Finally, a city railway is planned to connect the present central station (Porta Nuova) with the main station for high-speed trains (Porta Susa).

Sports Infrastructure

As a result of the Olympic Games, the sports infrastructure of Turin will enjoy several new facilities. Turin and the other towns involved in the organisation of the Games are committed in the coming years to taking care of the environment-friendliness of the new infrastructure environment – in line with the recommendations of Agenda 21 – and to guarantee their use after the Olympics, thus consolidating the development of tourism and providing a range of high-quality facilities. According to the green-card document (which analyses the impact of the Games on the urban environment) venues should meet three requirements:

1   to comply with public needs, regardless of the event for which they were designed;
2   to have a subsequent use that can be proved from real data and be backed by re-use programmes;
3   to keep the burden of their financial management acceptable for the territory in question.

Inspired by Barcelona, Turin wants to meet these requirements by developing multipurpose facilities. For instance, the speed-skating venue (Palasport Velocità) will become available to the city for indoor sports events (athletics), trade fairs, shows and exhibitions. Some of the spectator stands will be dismantled. These adjustments are needed because speed-skating is not a very popular sport in Italy. The short-track and figure-skating venue (Pavela) will be transformed into a centre for sports events, exhibitions and trade fairs. The new ice-hockey venue (Palasport Continassa) will also be used as a multipurpose sports centre.

Hotels and Tourism

A positive effect of the event may be the construction of new hotels. Currently, the city lacks a five-star hotel (as depicted in Table 6.2), and additional number of one- and two-star hotels are needed if Turin wants to attract leisure tourists. Most business hotels are occupied during the workweek and empty at weekends. Some recent investments to attract weekend guests (by reduced prices for instance) have achieved a slight rise in occupancy rates. The local government is aware of the need for more leisure-tourist hotels, but their construction is not included in the Olympic bid-book. According to this document, the city has drawn up plans to construct one new five-star hotel with 395 rooms and seven four- to five-star hotels with a total of 788 rooms.

Table 6.2   Existing hotels in Turin (city and region)

	City		Region	
	Hotels	Rooms	Hotels	Rooms
5 star	0	0	1	172
4 star	14	1,532	86	7,905
3 star	58	1,890	483	13,183
2 star	24	479	387	6,455
1 star	45	903	469	7,895
Total	141	4,804	1,426	35,610

Source: *Torino 2006* (Associazione Torino 2006, 1998).

Financing

The capital investments described above are not included in the budget of the organising committee (the OCOG budget). Table 6.3 presents an overview of such investments to be made by the local and national governments and by the private sector.

Table 6.3   Capital investments in millions of US$

Airport	141
Roads and railways	358
Visitor accommodation	198
Sport venues	374
Olympic village	76
Media	147
Total	1,294

Source: *Torino 2006* (Associazione Torino 2006, 1998).

Marketing: Events and Promotion

Several events have been planned to highlight the relation between sports and culture. According to the candidacy file, the city wants to host conventions, theatre festivals, concerts, films and exhibitions in the three years before the Olympic Games. These pre-Olympic events are meant to illustrate various aspects of winter sports combined with the economic and cultural activities of the local tradition (Associazione Torino 2006, 1998). The City Council aims to use the Olympic Games and the related cultural events as instruments to change the image of the city, by promoting Turin as a city of sports and culture. Furthermore, the city attaches great importance to the participation of the citizens. To that end, a special youth programme (which started in 2000) has been developed: a training programme for 15,000 volunteers who are prepared for the event by education in the fields of sport, culture and language. The programme is regarded not only as a necessity for staging the event, but also as a way to increase participation in sports.

The assignment of the Olympic Games has stimulated Turin to develop a marketing strategy. One of the objectives of that strategy is to change the image of Turin. From research, people appear to associate Turin with FIAT, Juventus and the Museum for Egyptian Art. When asked to mention two other characteristics of Turin, most interviewees have no reply. The conclusion must be that most people have a very incomplete perception of the city.

> The Olympics are indeed a powerful tool for promoting our region: the two weeks of sports events will put Torino in the international limelight and thousands of sports fans will be guests in our hotels, eating in our restaurants, travelling on our roads and using our services. Thousands of journalists from all around the world, and not only sports writers, will talk about Torino, assessing its efficiency and organisational capacity, its potential for the future, and its cultural and tourist attractions. Torino will be able to use the Winter Olympics to show itself as a welcoming hospitable location for economic, sports and cultural events, a place worth visiting before and after the event, where life and work are both pleasant and stimulating (City of Turin, 2000b, p. 3).

Turismo Torino – founded in 1998 – aims to market and promote Turin and its metropolitan area, providing the structures (information centres), tools (information and promotional materials) and activities that help convey a new image for the city towards (potential) visitors. Turin wants to enrich its image with new elements, as the city realises that improving the (tourist) product is not sufficient. Target groups of that communication strategy are international opinion-makers and opinion-leaders, including journalists. Turismo Torino emphasises three characteristics of Turin: 1) it is the capital of industry; 2) it is a romantic city; and 3) it is the capital of innovation (in the field of information and communication technologies). The organisation has deliberately decided not to promote Turin as a city of art (although the supply of cultural facilities is considerable), because of the heavy competition from other Italian cities in that respect. Recently, a new organisation – the Convention Bureau – was established to promote Turin as a destination for conventions.

Promotional activities are also undertaken by an international organisation which has a seat in Turin. This organisation is responsible for promoting events and the production of promotional brochures and videos, as well a bimonthly newsletter, which will be sent to opinion-makers and leaders throughout Europe. Another communication instrument is the road show: a tour around European cities where attractions in the realms of culture and sports can be presented. The Olympic Games have no key role in all these initiatives, but the event is certainly a suitable occasion to promote the city.

The Ambitions of Juventus

Turin is the home base of one of Europe's most successful football clubs, Juventus. Juve (as it is known) has won the European Champions Cup (Champions League) twice. As mentioned earlier, Juventus is one of the three main elements in the image of Turin, the others being FIAT and the Museum for Egyptian Art. Their team plays its home matches in Delle Alpi, a stadium with 69,000 seats built in the late 1980s to stage the 1990 World Football Championships. Although the stadium is quite impressive from an architectural point of view, many people complain about its qualities as a football stadium. Because of the athletics track, the distance between the spectators and the pitch is relatively great, which (according to some people) has negative implications for the ambience in the venue.

Currently, Juventus and the City Council are debating on the future of Juventus within or outside the city borders. Juve wants its own stadium, just like Torino Football Club has. To own a stadium is increasingly important for clubs with ambitions to a stock exchange quotation. Moreover, the club wants a training centre and facilities to accommodate the team's various needs and requirements. The club has proposed to demolish the Delle Alpi Stadium and build a new one on a smaller scale, surrounded by a commercial area of 100,000 m^2. In May 1995, Juve asked for definite guarantees for the future of the Delle Alpi Stadium. 'Should the project for demolishing the stadium and the building of a new ground be turned down, the club could find itself forced to move from Turin, the city where Juve has played for 100 years.'[1] There would be opportunity to buy a satellite football team from Trieste. A new stadium could be constructed at several locations. One of the potential locations is adjacent to Delle Alpi. Another option is the site of Stadio Comunale.

After months of discussion, the City Council has offered to give permission for the construction of a new stadium. However, the City Council objects to the size of the commercial area planned to surround the new venue. During the negotiations, the council offered to compromise on a size of 17,000 m². The City Council wants to limit the size of the commercial area in order to reduce the competition between that area and other commercial areas in the city. Juventus prefers a large commercial area to attract additional income, regardless of its location, within or outside the city borders. One municipality in the Turin region has already offered a site to Juventus.

Sports for All

The city of Turin actively stimulates participation in sports. One reason for doing so is the positive relationship between sports and health. To study that relationship, the City Council commissioned the regional office of the Sports Medicine Institute – responsible for monitoring the health of members of sports clubs – to examine all children between the ages of 10 and 11. That study and other investigations have revealed that the physical capacities of children in Turin have declined compared to 20 years ago, while the percentage of youngsters that participate in sports clubs has remained the same. Therefore, the physical problems are ascribed to a change in children's lifestyle, which is now dominated by computers. The Sports Medicine Institute argues that schools should activate children by paying more attention to physical education. The institute is trying to persuade the national government to invest more in physical attention, claiming that such investment reduces the costs of health care (now and in the future). The national government has developed a programme to increase participation is sports, but has handed over its implementation to the municipalities. From lack of financial means, the City Council looked for and found a company to sponsor the project: Robe di Kappa.

Robe di Kappa

Robe di Kappa started as a company in 1916. From the 1960s onwards, sponsoring became more and more important as competition from Asian countries grew. Robe di Kappa (a brand at that time owned by Maglifico Calzifico Torinese) survived that competition by using the power of sponsoring. From the late 1970s until 1994, Kappa derived its success from sponsoring the United States athletics team. In 1994, however, Kappa was declared bankrupt. The trademarks Robe di Kappa and Kappa were taken over by a new company – Basic Group – with a more decentralised organisational structure. The sales activities have been transferred to independent distributors ('licensees'). Although a global company, Basic Group is very committed to Turin. Robe di Kappa is the shirt sponsor not only of Juventus, but also of the city's most successful volleyball team, CUS Torino. Basic Group has decided to stay in Turin, while many other companies have left the city to find a cheaper location. Although production has been outsourced to companies all over the world, the old factory in Turin still accommodates the research, design and marketing activities, employing approximately 200 people.

The comprehensive programme to relaunch the company and its trademarks includes the renovation of the historical factory premises and the creation of Basic Village. This village accommodates a store (a so-called factory outlet), a supermarket, a restaurant and a travel agency. Other facilities to be opened soon are a bank, a hairdresser, a playground and a bar. Consequently, the factory and its surrounding area are slowly transforming into a visitor-attracting location. Kappa has an intensive relationship with the City Council and sponsors many urban projects and events, including projects to stimulate sports participation.

Projects

Several projects have been developed to stimulate (young) people to participate in sports. In 2000, for instance, one of the four national car-free Sundays was dedicated to sports. In various Italian cities, walking festivals were organised, combining sports and music. Another initiative is 'Open Doors', an event based on the 'sports for all' ('sport per tutti') philosophy. This is an annual event, organised by 105 municipalities in the Turin region. On that particular day, everyone is free to enter sports facilities and practise the sport of his or her choice, free of charge. In 1999, the event drew 70,000 interested people. At least 3,000 of them decided to practise the sport in question on a regular basis, to the satisfaction of the sport clubs.

Open Doors has been turned into an annual event, at least until the Olympic Games. The municipalities even decided to organise a winter edition, which took place for the first time in February 2000. Open Doors is a good preparation for the youngsters who have volunteered to participate in the organisation of the Olympic Games. The organising committee of Open Doors consists of representatives of all municipalities in the region. The event fits well in the strategy of the Turin province to upgrade its area, using sports as an instrument to promote the area and attract tourists. The province stimulates municipalities to invest in sports, for instance by constructing cycle lanes and footpaths, which opens opportunities to practise 'on-the-street' sports such as jogging, cycling and skating.

Conclusions

The assignment of the Olympic Winter Games to Turin has offered the city a great opportunity to invest in its attractiveness and its international image. Turin regards the event as a stimulus for the revitalisation of the city, which is outlined in a strategic plan.

The Games are expected to accelerate the implementation of infrastructural projects that improve internal accessibility not only within the city but also within the Piemonte region. By improving the connection between the town and the ski resorts, synergies between the complementary centres – in terms of tourism for instance – could be realised. Investment in local public transport (a city-rail connection) may help to reduce the air pollution caused by cars in the city. Furthermore, the event will result in the construction of new hotels, strengthening Turin's complementary tourist product. In the candidacy file the organising committee proposes building four- and five- star hotels, which may help to promote the city as a conference destination. However, to stimulate leisure tourism, the city should also stimulate the construction of affordable family hotels.

Turin will use the Olympics also to upgrade and expand its *regional* sports product, stretching from the mountains to the city. The city has adopted a sustainable approach, in which long-term interests play an essential role. In Italy however, the Olympic Winter Games do not seem to be an ideal event to construct facilities that can go on being used afterwards, as can be illustrated by the construction of an indoor speed-skating venue. Apart from skiing, the local population is not very active in winter sports. Nevertheless, the new venues will give the city the opportunity to attract sports events and hence to strengthen its sports image (branding). In general, the Olympic Games can endow the city with a competitive advantage over other Italian cities in sports (Rome being the only other Italian *city* with Olympic experience).[2]

The Olympic Games also offer the city a great chance to enrich its general image, which is now dominated by just a few elements. However, it cannot be concluded from the strategy of Turisme Torino that the city is planning to promote itself as a sports city (brand marketing). Instead, the Olympic games are regarded as one instrument to promote other features than sports (capital of industry, romantic city and capital of innovation). International media constitute the direct target group of this strategy, but indirectly the city tries to reach potential residents, visitors and companies (external target groups). In the strategic plan, the city aims to use sports as an instrument to attract specific target groups: participants and spectators. The city of Turin also considers the event to be a stimulus for participation in sports by the local population.

An interesting aspect of the Turin case is the role of companies in the field of sports and city marketing, where two extremities can be observed. FC Juventus is not very aware of its relationship with its surroundings. The club is even considering leaving the city to realise its ambitious plans. If the club decides to move to a suburb of Turin, the city could still benefit from the club's marketing potential. In that case, however, cooperation between the municipality and the club will be more complicated than ever, limiting the possibilities of using the club for city marketing. If the club decides to move away from the Turin region, the city will lose an essential element of its sports product. Despite the evident importance of Juventus for the city, the club is not included in the vision and strategy of the city. In the current discussions between the city and the club, both actors need to understand their common interests. Ideally that should result in an modified plan that benefits the club (by generating additional income for the club) as well as the city (by generating publicity and employment). Preferably, the commercial area around a new Juventus stadium should become a complement to rather than a substitute for existing commercial centres.

Basic Group (Robe di Kappa) is an example of a company that is deeply rooted in the city. This firm sponsors not only two clubs from Turin (including Juventus) but also a municipal programme (which is part of a national programme) to stimulate people to participate in sports. In contrast with the proposed Juventus arena, Basic Village is a complementary sub-centre with mainly sports-oriented activities.

Notes

1   Source: www.juventus.com.
2   Rome hosted the Olympic Summer Games in 1960. The Olympic Winter Games of 1956 were held in Cortina d'Ampezzo, a *village* (not a city) in northeast Italy.

Chapter Seven

Synthesis

Introduction

In the research framework of this investigation – presented in Chapter One – we presumed that sports have a growing impact on cities. In that vein of thought, an analysis of the opportunities to use sports as an instrument for city marketing seemed interesting to us. In the case studies of Barcelona, Helsinki, Manchester, Rotterdam and Turin (Chapters Two to Six), we have described the principal developments of sports and city marketing in these cities. Each of the cities can be characterised as sports-minded: Barcelona revitalised the city by means of the 1992 Olympic Summer Games. Helsinki has hosted several sports events, including the World Ice-Hockey Championships in 1997, and was one of the candidates for organising the 2006 Winter Games. Manchester will be the host of the Commonwealth Games in 2002, and Rotterdam was the Finals City of EURO 2000. The city of Turin has been designated to host the 2006 Olympic Winter Games.

In this chapter – the synthesis – we will confront the experiences of the five cities with the research framework. In each case study we have discussed the subjects introduced in the research framework: 1) sports and an attractive city; 2) the role of sports events, venues and clubs; 3) the use of sports in brand marketing and differentiated marketing; 4) the use of sports as catalyst for urban regeneration; and 5) the need for organising capacity. In practice, these matters appear to be closely connected, each of them being a potential element of an integrated sports and city marketing strategy. The synthesis will deal with each single element as well as with the connections among them, to reach some general conclusions from the five cases.

Sports and an Attractive City

The first theme that has been identified is the required awareness of the relation between sports and an attractive city. *Sports is important in the cities that have been analysed and, generally speaking, sports are regarded as an integral part of the city's attractiveness*. There is a growing awareness that sports cannot be treated separately from other urban policy areas. Barcelona was one of the first cities in Europe to find sports instrumental to other objectives of urban policy, including a raised quality of life. The experiences of Barcelona will be discussed in more detail in the next sections. Indeed, investment in sports, and in sports facilities in particular, can enhance the quality of life, which is such an important location factor nowadays.

Helsinki and Rotterdam are two cities that have paid much attention to the construction of sports venues. The budget of the Helsinki's municipal sports department for the construction and maintenance of sports venues equals that of the national government. In comparison with many other cities, the city of Helsinki boasts a broad range of sports opportunities and sports facilities, and subsidises sports halls, swimming pools and nearly 400 amateur sports clubs. In the course of time, Rotterdam has also developed a relatively generous supply of sports facilities both for top-class sports and for mass recreation. Another factor that sets Rotterdam apart is that the promotion of sports is delegated to two dedicated organisations: the separate foundation 'Rotterdam Topsport' focuses on top-class sports, and the Rotterdam Leisure Department deals with all other sports-related activities.

The case reports demonstrate that sports can help to integrate different groups in society and to educate youngsters. Moreover, sports can be considered a tool of a city's health policy. Youngsters appear to receive special attention in the sports policies of most cities. Together with the municipality, the Sports Medicine Institute in Turin has developed a programme to give sports a more prominent place in schools; it is sponsored and supported by the Turin-based Basic Group, producer of sportswear (one of its brands is Robe di Kappa). The case of Turin illustrates that sports are an economic activity in their own right. A study of the Manchester region estimates that the sport-related industry contributes 3 per cent to the regional economy, thanks to such well-known companies as Umbro, Reebok and a subsidiary of Adidas.

In spite of the increasing awareness of the role of sports as an integral part of an attractive city, the majority of the cases also demonstrate that the political priority for sports is not as high as one might expect, for instance in comparison with another important determinant of urban attractiveness, art and culture. In Turin, sports and culture come under the same alderman's responsibility, but the budget for culture is several times higher than that for sports. Although Helsinki shows the highest percentage of sports club memberships, cultural policies seem to have advanced ahead of sports policies. The resources and efforts that Rotterdam is spending on the preparation of the Cultural Capital of Europe 2001 project, considerably exceed those related to EURO 2000.

The first element of an integrated sports and city-marketing policy – sports as one of the urban qualities – is depicted in Figure 7.1.

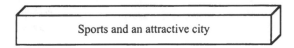

Figure 7.1   Awareness of the relation between sports and an attractive city

The Role of Sports Events, Venues and Clubs

As an integral part of an attractive city, sports require continuous investment in the city's sports product. In the first chapter, sports events, venues and clubs were identified as important features of that product. The events, venues and clubs naturally belong to the total supply of urban facilities and services, but in addition they could be instrumental to city marketing. In a discussion of sports as an instrument of city marketing several questions present themselves. What types of investment can be observed in the cities? What is the main goal of city-marketing efforts: to develop the city as a brand or to 'sell' the city to more specific target groups? Or do cities try to do both? And finally, what benefits to urban development do the cities expect from such investments?

All these subjects will be dealt with in succession. However, first we will review the most striking sports-related investments in the five cities (see Figure 7.2).

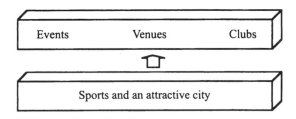

Figure 7.2   The role of sports events, venues and clubs (instruments)

Events

The hosting of major sports events is one of the most important opportunities for a meeting of sports and city marketing. That message was clear to all cities included in the comparative research. One classic example in that respect is the city of Barcelona, which turned the 1992 Olympic Games into more than just a sports event. The city is admired not only for the organisation of the Games but also for the way it used the event for the benefit of the city by means of a massive investment scheme. Events also played a considerable role in Manchester, given its investments in two Olympic bids (without success) and its successful bid for the 2002 Commonwealth Games. The latter event can be regarded as the main feature of Manchester's sports and city marketing policy, bringing a lot of additional resources to the city. From the moment EURO 2000 was allotted to the Netherlands and Belgium, Rotterdam has promoted its stadium (with five UEFA stars) as the ideal location for the finals. To the surprise of many, Rotterdam defeated the Dutch capital Amsterdam in the competition for these finals.

Inspired by the Olympic success of Barcelona, Turin decided to run for the 2006 Olympic Winter Games. In 1999, the IOC favoured the Italian city over Sion (Switzerland) and Helsinki. Turin now faces the challenge of living up to the promises made in the bidding process and being ready for the Games on time. Helsinki was a strong competitor to Turin in the bidding process, since many of the required sports facilities were already available there in 1999. The city has a track record in hosting major sports events such as the IAAF World Championships Athletics in 1983 and 1994 and the World Ice-Hockey Championships in 1997. The use of events for city marketing has played a role in every single case, as will be illustrated further on in this chapter.

Venues

The second type of investment relates to venues. In some cases, such investments have been made to stage a specific event; in others, venues have been developed independently. Each of the cities involved in our study invested recently in venues or is planning to do so. In the northern part of Helsinki's city centre, a state-of-the-art indoor arena has been constructed. Its principal financier is a private investor (the Jokerit HC Group); only 6 per cent of the investment is covered by the municipality. This multifunctional venue is provided with an advanced information and communication infrastructure – including optical wiring – as was to be expected in one of Europe's capitals in the field of ICT. Another facility built recently – with heavy financial involvement of the same private investor – is a new football stadium (Finnair Football Stadium).

Rotterdam has invested in the renewal of Feijenoord Stadium and two other football stadiums and in the construction of an indoor arena ('Topsportcentrum Stad Rotterdam Verzekeringen'), which opened its doors to the public in January 2000. This arena is complementary to other major accommodations such as Ahoy' Rotterdam and Feijenoord Stadium. The operators of these venues will put in their expertise to manage the arena.

Manchester has invested in the National Cycling Centre (Manchester Velodrome) and a new arena for basketball. To sustain its Olympic bid, these facilities were realised during the bidding process. The National Cycling Centre is located in Sportcity, an area of 40 hectares in the eastern part of Manchester, which will be the focal point of the 2002 Commonwealth Games. At the heart of this area, the new City of Manchester Stadium (the new home of the Manchester City Football Club) will be erected, along with some other new facilities. The investment scheme also includes a two-storey swimming pool within the university complex, south of Sportcity.

A New Look at Sports Venues: Barcelona Park

The case of Barcelona illustrates that a football stadium cannot be treated separately from its surroundings. FC Barcelona is not building a new stadium but has taken the initiative to develop a new urban sports product: Barcelona Park, which is planned in the direct environment of Nou Camp, the club's home base. The club presents Barcelona Park as an urban theme park with sports-related attractions, with FC Barcelona as its flagship. The trend to extend the control of a major sports club to its direct environment is also visible in Turin, where Juventus is eager to increase the spatial scale of its activities. The club has expressed the wish to leave its current stadium (Delle Alpi) for a new complex, preferably at the same site, surrounded by an area of 100,000 m^2 for commercial development.

Clubs

The examples of Barcelona and Turin touch upon the third important type of investment in sports and city marketing. European top clubs such as Manchester United, Juventus (Turin) and FC Barcelona could be the flagships of a sports and city marketing policy. To some extent, Feyenoord could play a similar role for the city of Rotterdam. Helsinki appears to lack such a trump card. Its most successful club – HJK – has only once qualified for the prestigious Champions League. *Surprisingly, the four cities with a European top football club do not give their club a prominent place in their sports and city marketing policy.*

The Use of Sports in Brand and Differentiated Marketing

As depicted in Figure 7.3, sports can be used for brand marketing and differentiated marketing.

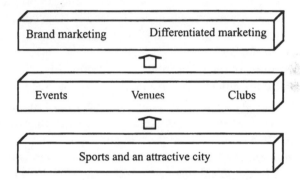

Figure 7.3   The use of sports in brand and differentiated marketing

Brand Marketing

The first chapter emphasised that the city can be seen as a 'brand with a broad supply of urban products'. Image and identity are important location factors, but cannot by themselves change the general perception of a city or a region. The brand marketing of a city has emerged as crucial from each of the five case studies. Major sports events are focal points of their brand-marketing strategies. In the comparative analysis, the city of Barcelona stands out from the other four cities.

Several city charts confirm Barcelona's popularity and strong positive image among tourists and business visitors. The city has succeeded in turning its image around by means of the 1992 Olympic Games. That event has been instrumental to many objectives of urban policy, including that of changing the city's image. Before, during and after the Games, the city profited from positive media coverage. Both the event and the urban renewal programme generated much free publicity, as is illustrated by the fact the number of journalists covering the event reached a world record. Nor did the media stop paying attention after the Games: they went on to cover the massive renewal projects. Obviously, Barcelona has established its own brand, and now finds itself in the luxurious position where other actors (such as travel agencies and the higher layers of government) invest in the continuity of Barcelona's 'brand'

The four other cities do not top the European city charts in terms of image and attractiveness. Manchester and Turin are trying to create an image that is no longer dominated by their industrial quarters and Rotterdam wants to be perceived as more than a port city. The peripheral location of Helsinki in Europe induces the city to try and put itself more firmly on the European map. For Helsinki, sports could be instrumental to outgrowing its peripheral image. The successful developments in the ICT sector (Nokia and other firms) have already improved the city's image, but these firms have still trouble recruiting international staff.

The conclusion from the investigation can be that brand marketing is an important stimulus for cities to run for and host major sports events. At the same time, the cities are aware that images are very hard to change and that modification takes a long time. Consequently, the effective use of sports events as a way to establish a positive image demands a long-term strategy, as is confirmed by Barcelona's Olympic experiences. The other four case studies show a variety of strategic approaches. In the case of Helsinki, the Olympic bid was not part of an explicit sports-events strategy or a general event strategy. Most of the city's energy has been spent on its status as one the nine Cultural Capitals of Europe 2000. Prior to the Olympic bid there was no long-term sports event strategy for the city of Turin. During the bidding process the city formulated a strategy incorporating many of the lessons from Barcelona and other cities. Now the strategy is being translated into an action plan.

Among the four cities that badly need to convert their image, Manchester has adopted the most explicit strategy. To Manchester, the participation in the bidding processes for sports events has been an objective in its own right. It started with the impulsive decision of an enthusiastic theatre director to run for the 1996 Olympics. Building upon that initiative, Manchester entered the bidding process again for the 2000 Olympics. That bid failed, but Manchester succeeded in getting the 2002 Commonwealth Games. Before the first Olympic bid Manchester was not regarded as a city of sports; now the city has gained a place in the world of sports and has already been host to some smaller (inter)national sports events.

Rotterdam has also entered the world of sports events. That the finals of EURO 2000 are integrated into a long-term sports-event strategy cannot be maintained, however. There is not enough budget and there is no strategy for the legacy of the EURO 2000 event. What is interesting, though, is that

Rotterdam created synergies between EURO 2000 and the Cultural Capital year 2001. Furthermore, a debate has started about the possibility of continuing the sequence of major events through a major maritime manifestation in 2003. In that scenario, cultural and sports events are linked. In fact, the same idea prevails in Barcelona, where the city, jointly with UNESCO, is staging another major cultural event named 'Forum of Cultures' for the year 2004, which could be a new instrument of urban revitalisation.

Sports events are a powerful tool for developing the city as a brand. Nevertheless, several dangers threaten the synergy between sports and city marketing. In the months before EURO 2000, the media and the national government paid most attention to the threat of hooliganism and not so much to the positive side of the event. Everybody was well aware that serious confrontations between rival supporters would be detrimental to the image of Rotterdam as a city of sports. With major sports events like the Commonwealth Games in Manchester and the Olympics in Turin, there is always the question of the allocation of the resources. Turin is at the opening stage of the project, when financial and media control is of the essence.

Sports venues have brand-marketing power as well, provided a special design, concept or location is involved. The development of innovative sports facilities gets a lot of attention within and outside the world of sports. The plans for FC Barcelona Park go beyond the plain and simple function of a football stadium. Already at the time of writing, Camp Nou is regarded as one of the city's top tourist attractions. The plan to combine a city park with special sports entertainment facilities is a new step forward in the development of sports venues. Clearly, the project has to overcome severe difficulties in adjusting to the surroundings. For one thing, the concept needs to be amended in terms of access, with a view to the projected tourist flows to the new theme park.

Finally, the brand-marketing power of European top football clubs that are named after their city or region is undisputed. However, even that element of the city's sports product has its reverse side: the reputation of hooliganism can rub off, as Feyenoord Rotterdam has experienced.

Differentiated Marketing

The investigation has confirmed that sports are important for city brand marketing. However, sports (events, accommodations and clubs) can also be used for more differentiated marketing efforts, directed to specific target groups. The distinction between brand marketing and differentiated marketing is important: investment in sports products can be oriented to specific groups, or particular segments of the market.

Most cities that stage a major sports event set great store by making the local community share its benefits. Two good examples of internal city marketing are provided by Manchester and Rotterdam. From day one the politicians in Rotterdam stressed that Rotterdam's old neighbourhoods should be allowed to enjoy the football championships. That approach is in line with the city's strategy of social revitalisation. Special neighbourhood programmes contain activities related to the championships (games, sports and culture). Externally, Rotterdam hoped that the spectators of the games would lengthen their stay in the city and become future visitors. An important group for targeted programmes was the international press, which by various means was supplied with all sorts of information on Rotterdam. The city also developed special programmes for supporters from countries whose clubs played a match in Rotterdam (the Portuguese, for instance).

For the Commonwealth games, Manchester has identified specific internal target groups, such as immigrants from Commonwealth countries. There is also a social regeneration programme which emphasises the involvement of the local community in the Social Volunteer Programme. Specific local small and medium-sized businesses should benefit from the commercial opportunities that the Games generate. Externally, certain types of business in the Commonwealth countries are targets for seminars, trade missions, etc.

In Barcelona, there has been a variety of programmes for visitors and investors since the international breakthrough of 1992, but they are not specifically sports-related. The city's tourist organisation has developed more of such dedicated programmes. International press offices are an important target for Helsinki's image-enhancing efforts. Most successful have been the endeavours promoting the city as one of Europe's leading technology cities.

Sports as Catalyst for Urban Regeneration

Interaction between sports and city marketing is best achieved when sports is the catalyst to an integrated urban regeneration process, as depicted in Figure 7.4.

The legacy of the Barcelona Games does not only concern new sports accommodations, visitors, and free publicity. The Games were the 'excuse' for the city to put in place a massive investment scheme which has accelerated the development of the city tremendously. It is the combination of an appealing

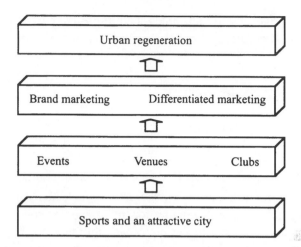

Figure 7.4 Sports as catalyst for urban regeneration

sports event with a massive urban-renewal project that made the Games extremely successful for Barcelona, and even now, more than 10 years on, still keeps the Olympic spirit alive in the city. Its effects should not be underestimated. The event induced a structural increase of urban tourism and boosted the city's attractiveness as a location for investors, businesses, visitors and of course the local community.

A closer look at the other cities investigated shows that Manchester has adopted a similar strategy. Manchester's 1995 sports strategy centres on events, facilities and participation. The Commonwealth Games provide an ideal opportunity to unite the three cornerstones of the strategy. The plans for the Commonwealth Games centre around that legacy. The event will take place in the eastern part of Manchester, one of the city's most deprived areas. The new sports facilities are under construction in Sportcity (an area of 40 acres in that part of the town). The legacy is concerned with new infrastructure as well as with the intangible effects of developing and strengthening networks with some of the Commonwealth countries (Malaysia, Singapore, Australia, Canada and South Africa). Manchester's objectives are defined in five categories: economic growth and employment; social regeneration ('the inclusive games'); sport-developing community participation; linking culture and the city to sports; and promoting Manchester and the northwest tegion.

In the course of time, Turin might also join the group of cities where sports has been a catalyst for urban regeneration. It has drawn up an ambitious programme in which the Games are the 'lever' to induce investment in neighbourhood redevelopment and the implementation of new public transport infrastructure. The legacy of the Games is an important part of the bid-book, but Turin is still at the stage of translating strategy into action.

The Need for Organising Capacity

A Growing Number of Stakeholders

Up to now, we have focused on the potential role of sports in city marketing as found in the five case studies. However, the synergy between sports and city marketing is not a matter of 'witchcraft'. In fact, many of the subjects that we have discussed are not a matter for the city government alone. In the field of sports and city marketing many stakeholders are involved, with different intentions, interests and resources. The trend is that the number of stakeholders is growing and the relations among stakeholders are getting more complicated. To create synergies in sports and city marketing makes high demands on the city's organising capacity. That is true of all levels of sports and city marketing, as depicted in Figure 7.5. Sports as an integral part of an attractive city already involves many stakeholders (politicians, municipal departments, sports clubs, sports facility managers, volunteer organisations, sponsors, etc.). An integrated and creative approach is required, such as the triangle formed by Basic Group (Robe di Kappa), the Sports Medicine Institute and the city of Turin to re-engineer sports in the local school system.

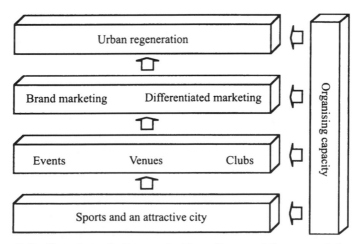

Figure 7.5   Sports and city marketing: the need for organising capacity

The more sports are explicitly connected to city marketing (brand marketing, differentiated marketing, as a catalyst for urban transformation), the more stakeholders are involved, which again challenges the city's organising capacity. From the many examples of such changing relations, we have selected three that have attracted our special attention: the need to consolidate the interests of cities, national governments, sports organisations and sports-marketing organisations to optimise the legacy of major sports events; the weak relationship between top-class professional football clubs and the city (global versus local); and the need for direction and supervision.

Major Sports Events and the Interests of Many Actors

Major sports events have always been a challenge to urban communities but the playing field is changing. On the one hand, globalisation and commercial-isation have drastically upset the world of major sports events. It is no longer a matter of just the IOC, the national sports organisations or the athletes: the influence of several intermediary agencies is waxing. These agencies buy the commercial rights to the events, as ISL did for the EURO 2000 football championships. They safeguard the interest of the official sponsors, to whom serious commitments have been made. ISL is a major player, but there are other sports-marketing agencies. On the other hand, the hosting nations and cities have raised their demands for hosting sports events in terms of international marketing, law and order and crowd control, their own sponsor programme, etc. Metaphorically, the world of sports events is a game where the rules become more complex and the number of players keeps increasing. The key challenge is to consolidate all the interests of these different players. The case of Rotterdam illustrates that all the stakeholders have a motive for reaching an agreement that reconciles the ambitions of cities and the sports world. The instrument of *strategic networking* is vital in such a setting and calls for a better understanding among the stakeholders through investment in the relations. Ideally, what is needed is a *joint vision* of the stakeholders on an integrated approach to sports events in cities. None of the stakeholders has an interest in such an outright confrontation as some cities in the Netherlands and Belgium have had with ISL/EURO 2000. It might cost the sports event itself the support of the local community.

Football Clubs: Global versus Local

At first sight, top European clubs such as Manchester United, FC Barcelona and Juventus would be expected to be a central feature of the sports and city marketing strategies of Manchester, Barcelona and Turin. We found, on the contrary, a troublesome relationship between the city administration and these clubs in all three cities. Globalisation and commercialisation have an impact on the professional football clubs as well. The clubs feel the pressure of international competition. In response to these trends, clubs seek to improve their financial position in many ways. For example, Manchester United is a company quoted on the stock exchange and their merchandising business has expanded considerably. The latest trend is that clubs hope to increase their budget through real estate development, as illustrated by FC Barcelona and Juventus. Consequently, the globalisation and commercialisation of the football business induces these clubs to go into real-estate development, with a direct impact on the city's development. This trend increases the pressure on the relationship between clubs and the city administration. While explanations vary, the troublesome relationship is counterproductive in all three towns. In the case of Manchester United, the location of the stadium just outside the administrative borders of the city (in the borough of Trafford), makes the situation more difficult for both sides. Despite the administrative problems, no-one can deny that Manchester United is one of the strongest brands in professional football, with an international fan base and goodwill dating back to the era of Matt Busby's 'babes'. The club does play a part in promotional videos and brochures and a closer working relationship with Marketing Manchester is developing. An improved relationship between the city and the club is vital. The relations between FC Barcelona and the city are not happy either. FC Barcelona differs from Manchester United, which is a company quoted on the stock exchange. With Barcelona, it is the members (more than 100,000 *socios*) who decide on the club's principal policy matters, so that Barça does have strong ties with the local community. Nevertheless, the matter of the Barcelona Park project has revealed a gap between the club and the public authorities in general. The innovative project does not automatically fit in its direct environment. Again, the city, the region, the opponents from the neighbourhood and the club should join forces to come up with a plan that is acceptable to all. The relationship between club and city is most troublesome in Turin, where Juventus has threatened to leave Delle Alpi and the city for the suburbs or the city of Triest. The Juventus' claim for additional land for commercial development is not based on a concept like FC Barcelona Park, which could become one of the city's top tourist attractions. Nevertheless, continued disagreement is good for neither Turin nor the reputation of Juventus.

Sports and City Marketing: The Need for Direction and Supervision

A final theme to be addressed is leadership. Complex urban revitalisation projects call for strong and creative leadership. The field of sports and city marketing clearly needs direction and supervision as well. Some cities have intensified their city-marketing efforts. Rotterdam for one has established a special organisation, 'Rotterdam Marketing', to orchestrate and coordinate all the somewhat isolated efforts to market the city. Sports and city marketing could be included as well, in partnership with such specialist organisations as 'Rotterdam Topsport'. Since its establishment, Marketing Manchester has developed a similar role for the city of Manchester, making sports and city marketing one of their policy fields. The need for direction and supervision is also very clear as regards the major sports events. The special organisational structure of the Olympic Games in Barcelona is an illustration of that need.

Sports and City Marketing in Perspective

The globalisation and commercialisation of the world of sports have greatly changed the impact of events, venues and clubs on city marketing. Professional football clubs are developing into powerful economic agents, sports events are first-class sponsoring opportunities for the business community, and sports venues have become a commercially-interesting meeting place. Many of the stakeholders involved are not from the city itself but operate on an international or even global scale. Their interests are sometimes contradictory to the concerns of the city administration and general societal interests.

The stakeholders – city councils, city departments, the management of sports venues and sports clubs, sports marketeers, licensees and the media – operate too independently. These players in the field of sports and city marketing need to be aware that they can be seen as part of a sports cluster, with potentially strong mutual interests. That calls for a change in the attitude of the institutions in the 'world of sports' on the one hand and the city on the other. The challenge is to develop strategic cooperation between the actors in the sports cluster in order to generate substantial synergies that can be beneficial to all actors, including clubs and cities. Knowledge and expertise in the area of 'sports and city marketing' are a means to raise the awareness among these actors. To cope with the ever-increasing complexity of the relationship between sports and city marketing, continued investment in the development of knowledge and expertise in that field is an absolute necessity.

References

Research Framework

Berg, L. van den (1987), *Urban Systems in a Dynamic Society*, Gower Publishing Company, Aldershot.

Berg, L. van den and E. Braun (1998), *From Gelredome to a Gelredome complex*, Euricur Report, Rotterdam.

Berg, L. van den, E. Braun and J. van der Meer (1997), *Metropolitan Organising Capacity; Experiences with Organising Major Projects in European Cities*, Ashgate, Aldershot.

Berg, L. van den, R. Drewett, L.H. Klaassen, A. Rossi and C.H.T Vijverberg (1982), *Urban Europe, a Study of Growth and Decline*, Pergamon, Oxford.

Berg, L. van den, L.H. Klaassen and J. van der Meer (1990), *Strategische City-Marketing*, Bedrijfskundige signalementen 90/3, Academic Service, Schoonhoven.

Berg, L. van den, J. van den Borg and J. van der Meer (1995), *Urban Tourism*, Avebury, Aldershot.

Bramezza, I. (1996), *The Competitiveness of the European City and the Role of Urban Management in Improving the City's Performance*, Tinbergen Institute Research Series, No. 109, Erasmus University, Rotterdam.

Castells, M. (1989), *The Informational City: Information Technology, Economic Restructuring and the Urban-Regional Process*, Blackwell, Oxford.

Castells, M. (1996), *The Information Age: Economy, Society and Culture*, Vol. I, *The Rise of the Network Society*, Blackwell Publishers, Oxford.

Duncan, T. and S. Moriarty (1997), *Driving brand value using integrated marketing to manage profitable stakeholder relationships*, McGraw-Hill, New York.

European Commission (1991), *De Europese Gemeenschap en Sport: mededeling van de Commissie aan de Raad en het Europees Parlement*, Brussels.

Faure, J.M. (1994), 'Sport: vermaak voor de massa en cultuur voor de elite', in W. Blockmans (ed.), *Europa door de eeuwen heen; Wetenschap, Transport, Oorlogen, Sport & Spel*, Gezondheid en Kunst, Kosmos-Z&K, Utrecht/Antwerp.

Flicke, F. (1999), 'Milliardenmarkt Sport-business', *Bizz*, 1 February.

Hall, P. and D. Hay (1980), *Growth Centres in the European Urban System*, Heinemann, London.

Hall, P. (1995), 'Towards a General Urban Theory', in J. Brotchie, M. Batty, E. Blakely, P.Hall and P. Newton (eds), *Cities in Competition: Productive and Sustainable Cities for the 21st Century*, Longman Australia, Melbourne.

Serail, S. (1993), *Sport uit de marge: ervaringen uit tien grote steden*, Ministerie van Welzijn, Volksgezondheid en Cultuur, Directie Sportzaken, Sport en Sociale Vernieuwing 7, Rijswijk.

Barcelona

FC Barcelona (1999), *Parc del Barça; The Project.*

Guevara, L.M. de, X. Còller and D. Romanì (1995), 'The Image of Barcelona '92 in the International Press', in M. de Moragas and M. Botella (ed.) (1995), *The Keys to Success: The Social, Sporting, Economic and Communications Impact of Barcelona '92*, Centre d'Estudis Olímpics i de l'Esport, Universtat Autònoma de Barcelona.

Nel.lo, O. (1998), 'El canvi social a la regió metropolitana de Barcelona: deu preguntes', in *Dossier: La Barcelona metropolitana: economia i planejament*, Col.legi de'Economistes de Catalunya, no. 34.

Turisme de Barcelona (1998), *Tourism Statistics 1998*.

Helsinki

2006 Association (1998), *Helsinki 2006: Candidate City; Olympic Winter Games Candidature File*.

Helsinki City Council (1999), *New Policy Priorities for Helsinki's International Activities*.

Jokerit HC Group (1999), *Hartwall Areena*.

Urban Facts (1997), *Helsinki 1997; Facts about Helsinki*, City of Helsinki.

Uusimaa Regional Council (n.d.), *Helsinki Region: A European Province with Prospects and Potential*.

Manchester

City Pride Partnership (1994), *City Pride: a Focus for the Future*.

City Pride Partnership (1997), *City Pride 2: Partnerships for a Successful Future*.

House of Commons: National Heritage Committee (1995), *Bids to Stage International Sporting Events*, Session 1994–95, Fifth Report, HC 493, HMSO, London.

Manchester 2002 Ltd (1998), *Manchester 2002 News*, issue 2, December.

Manchester Leisure and Manchester Education Services (1999), *Towards the Commonwealth Games; Manchester 2002*.

Office for National Statistics (1995), 'AES', in Manchester TEC (1998), *Economic Assessment 1998*.

Phelan, L. (1997), 'Manchester: A City on the Move; Sports as a Tool for City Marketing', paper presented to the ICSS workshop on the economic impact of sport, International Centre for Sports Studies, Neuchatel, Switzerland.

Sports England (1999), *The Value of Sport: Best Value through Sport*.

Strategic Leisure, Chesterton and Sheffield Hallam University (1999), *Sports-related Industry*, MIDAS, Manchester City Council and The North West Development Agency.

The 2002 North West Partnership (1999), *The 2002 NW Economic and Social Programme 1999–2004: A Bid to the SRB Challenge Fund*.

Rotterdam

City of Rotterdam (1998), *Met Raad en Daad: Collegeprogramma 1998–2002*.
Project Bureau Rotterdam EK 2000 (1999), *Masterplan Rotterdam EK 2000*.
Rotterdam City Development Corporation (RCDC) (1998), *Trendbericht Rotterdam 1998: Sociaal-economische rapportage*.
Rotterdam Leisure Department (1998), *Stand van zaken en vooruitblik*, April 1998.
Rotterdam Leisure Department (1999), *Inventory of Sports Facilities in Rotterdam 16–3–1999*.
Van der Vegt, J. (1998), *Rotterdam Orgware*, Rotterdam City Development Corporation.

Turin

Associazione Torino 2006 (1998a), *Torino 2006*, official candidature file.
Associazione Torino 2006 (1998b), *Green Card Torino 2006*.
Borg, J. van den, A.P. Russo and G. Rumi (1999), *The Visitor-friendly Metropolis: An International Comparative Investigation into the Hospitality Offered to Visitors of Birmingham, Lisbon, Lyons, Nantes, Rotterdam and Turin*, Euricur Report, Rotterdam.
City of Turin (2000a), *Torino Internazionale: Piano strategico per la promozione della Città*.
City of Turin (2000b), *Torino: A City to Discover*, international newsletter, no. 1, February/ March.

Appendix: Discussion Partners

Barcelona

A. Batlle i Bastardas, Deputy Major of Sports and Institutional Relationship, City of Barcelona

J. Coll, General Director, FC Barcelona Foundation

J.M. Costa, Manager Project 2000, Football Club Barcelona

I. de Delàs, Director of Planning and Control of the Administration, Turisme de Barcelona

P. Duran, Director, Turisme de Barcelona

R.M. Fraile, General Director, Turisme de Barcelona

A. Gil-Vernet i Huguet, Advocats Associats, Representative of Association against FC Barcelona park

M. Ibern, former Head of the Sports Department, City of Barcelona

A. Junceda, Commercial Director, FC Barcelona Foundation

M. de Moragas Spà, Institute of Communication, Autonomous University Barcelona

C.M. Muntañola, General Director, Real Club de Tenis Barcelona

J.M.T. Turull, President of Real Club de Tenis Barcelona

A.B.M. Vílchez, Centre for Studies on Olympics and Sports, Autonomous University Barcelona

Helsinki

A. Bryggare, member of the City Council, winner of Olympic medal

C. Forsell, Marketing Project Manager, Football Association of Finland

O. Lahtinen, Deputy Tourist Director, Tourist Office, City of Helsinki

P. Mustonen, Head of the Information Bureau, City of Helsinki

J. Piirainen, Secretary General, Finnish Sports Federation

A. Rauramo, Head of Sports Department, City of Helsinki

M. Sulin, General Manager, Jokerit HC Group

M. Vanni, Managing Director, Event Partnership Finland

A. Viinikaa, Deputy Major, City of Helsinki

L. Wangel, National Team Director, Football Association of Finland

Manchester

N. Allen, Principal Consultant, Strategic Leisure Ltd
D. Carter, Manchester City Council
C. Gratton, Director, Leisure Industries Research Centre, Sheffield Hallam University
C. Jones, Senior Consultant Planning and Economics, Chesterton International Property Consultants
P. Knowles, Head of Sports Department, City of Manchester
D. Nicholas, Director of Planning and Economics, Chesterton International Property Consultants
L. Phelan, Deputy Chief Executive, Marketing Manchester
J. Quigley, Manchester City Council

Rotterdam

H. de Bruin, Rotterdam Municipal Port Management
W. Buitendijk, Spo Mark, co-organiser of the ABN AMRO World Tennis Tournament and the Concours Hippique International Officièl (CHIO)
W.F. Huibregsten, former President, NOC*NSF
B. Lenstra, Director, B-producties
J. van Merwijk, Director, Stadion Feijenoord NV
J. Moerman, Director, Rotterdam Festivals Foundation
J.H.A. van der Muijsenberg, Alderman, Sports and Recreation, City of Rotterdam
J. Noordenbos, former Director, Rotterdam Topsport Foundation
W. Noordzij, Municipal Project Manager, EURO 2000
H. den Oudendammer, Director Rotterdam Topsport Foundation
G. Reussink, Director, Rotterdam Leisure Department, City of Rotterdam
J. van der Vegt, Director, Ahoy' Rotterdam NV
J. van 't Verlaat, Rotterdam City Development Corporation
H. de Wilde, Marketing Manager, Professional Football Feyenoord Foundation
H. Zoethoutmaar, Project Manager, Rotterdam Topsport Foundation

Turin

P. Bellino, City of Turin
M. Berruto, Head Coach, A.S. Pallavolo (Volleyball) Torino
A. Carbonara, Marketing and Communications Director, BasicNet
L. Chiabrera, Director of Sports Facilities and Institution Relations, Torino
  2006 Bid Committee/President Turin Marathon
M. Damilano, Marketing Advertising Promotion Italia
R. d'Elicio, National Manager of Athletics, University Sports Centre Turin
E. Ferro. President of the Regional Committee of Piemonte
R. Gilardo, Turin Marathon
C.G. Gribaudo, Director of the Turin Sports Medicine Institute
A. Ippolito, Sports Director, A.S. Pallavolo (Volleyball) Torino
U. Perrone, City of Turin
F. Soncini, City of Turin

Index

Printed and bound by CPI Group (UK) Ltd, Croydon, CR0 4YY

21/10/2024

01777082-0012